MW00845011

Liquid Biopsy

Liquid Biopsy
New Challenges in the Era of Immunotherapy and Precision Oncology

Edited by

ANTONIO RUSSO
Department of Surgical, Oncological, and Oral Sciences, University of Palermo, Palermo, Italy

ETTORE CAPOLUONGO
Department of Molecular Medicine and Medical Biotechnology, University of Naples Federico II, Naples, Italy
CEINGE, Advances Biotecnologies, Naples, Italy

ANTONIO GALVANO
Department of Surgical, Oncological, and Oral Sciences, University of Palermo, Palermo, Italy

ANTONIO GIORDANO
Sbarro Institute for Cancer Research and Molecular Medicine, and Center of Biotechnology, College of Science and Technology, Temple University, Philadelphia, PA, United States

LIBRARY OF CONGRESS
SURPLUS
DUPLICATE

ACADEMIC PRESS
An imprint of Elsevier

Academic Press is an imprint of Elsevier
125 London Wall, London EC2Y 5AS, United Kingdom
525 B Street, Suite 1650, San Diego, CA 92101, United States
50 Hampshire Street, 5th Floor, Cambridge, MA 02139, United States
The Boulevard, Langford Lane, Kidlington, Oxford OX5 1GB, United Kingdom

Copyright © 2023 Elsevier Inc. All rights reserved.

No part of this publication may be reproduced or transmitted in any form or by any means, electronic or mechanical, including photocopying, recording, or any information storage and retrieval system, without permission in writing from the publisher. Details on how to seek permission, further information about the Publisher's permissions policies and our arrangements with organizations such as the Copyright Clearance Center and the Copyright Licensing Agency, can be found at our website: www.elsevier.com/permissions.

This book and the individual contributions contained in it are protected under copyright by the Publisher (other than as may be noted herein).

Notices
Knowledge and best practice in this field are constantly changing. As new research and experience broaden our understanding, changes in research methods, professional practices, or medical treatment may become necessary.

Practitioners and researchers must always rely on their own experience and knowledge in evaluating and using any information, methods, compounds, or experiments described herein. In using such information or methods they should be mindful of their own safety and the safety of others, including parties for whom they have a professional responsibility.

To the fullest extent of the law, neither the Publisher nor the authors, contributors, or editors, assume any liability for any injury and/or damage to persons or property as a matter of products liability, negligence or otherwise, or from any use or operation of any methods, products, instructions, or ideas contained in the material herein.

ISBN: 978-0-12-822703-9

For Information on all Academic Press publications
visit our website at https://www.elsevier.com/books-and-journals

Publisher: Stacy Masucci
Acquisitions Editor: Linda Versteeg-Buschman
Editorial Project Manager: Susan E. Ikeda
Production Project Manager: Maria Bernard
Cover Designer: Vicky Pearson Esser

Typeset by MPS Limited, Chennai, India

Contents

6.6 Cell-free/circulating tumor DNA profiling: from next-generation sequencing-based to digital polymerase chain reaction-based methods

A. Perez, C. Brando, M. La Mantia, V. Gristina, A. Galvano, L. Incorvaia, G. Badalamenti, A. Giordano, E. Capoluongo, A. Russo, U. Malapelle and V. Bazan

6.7 Standardization and quality assurance in liquid biopsy testing

A. Perez, M. Bono, N. Barraco, C. Brando, D. Cancelliere, A. Pivetti, A. Fiorino, E. Pedone, A. Giurintano, M. La Mantia, V. Gristina, A. Galvano, L. Incorvaia, G. Badalamenti, U. Malapelle, A. Russo and V. Bazan

7. Early detection screening: myth or reality?

M. La Mantia, F. Iacono, S. Cutaia, V. Gristina, A. Perez, M. Greco, K. Calcara, A. Galvano, V. Bazan and A. Russo

List of contributors

M. Arbitrio
Institute for Research and Biomedical Innovation (IRIB), Italian National Council (CNR), Catanzaro, Italy

G. Badalamenti
Department of Surgical, Oncological, and Oral Sciences, University of Palermo, Palermo, Italy

N. Barraco
Department of Surgical, Oncological, and Oral Sciences, University of Palermo, Palermo, Italy

V. Bazan
Department of Biomedicine, Neuroscience and Advanced Diagnostics (Bi.N.D.), University of Palermo, Palermo, Italy

M. Bono
Department of Surgical, Oncological, and Oral Sciences, University of Palermo, Palermo, Italy

C. Brando
Department of Surgical, Oncological, and Oral Sciences, University of Palermo, Palermo, Italy

G. Busuito
Department of Surgical, Oncological, and Oral Sciences, University of Palermo, Palermo, Italy

F. Buttitta
Center of Advanced Studies and Technology - University of Chieti, Chieti, Italy

K. Calcara
Department of Surgical, Oncological, and Oral Sciences, University of Palermo, Palermo, Italy

D. Cancelliere
Department of Surgical, Oncological, and Oral Sciences, University of Palermo, Palermo, Italy

E. Capoluongo
Department of Molecular Medicine and Medical Biotechnology, University of Naples Federico II, Naples, Italy; CEINGE, Advances Biotecnologies, Naples, Italy

D. Caracciolo
Department of Clinical and Experimental Medicine, Magna Graecia University of Catanzaro, Catanzaro, Italy

M. Castiglia
Department of Surgical, Oncological, and Oral Sciences, University of Palermo, Palermo, Italy

A. Cordua
Department of Clinical and Experimental Medicine, Magna Graecia University of Catanzaro, Catanzaro, Italy

O. Cuomo
Department of Clinical and Experimental Medicine, Magna Graecia University of Catanzaro, Catanzaro, Italy

S. Cusenza
Department of Surgical, Oncological, and Oral Sciences, University of Palermo, Palermo, Italy

S. Cutaia
Department of Surgical, Oncological, and Oral Sciences, University of Palermo, Palermo, Italy

M. Del Re
Unit of Clinical Pharmacology and Pharmacogenetics, Department of Clinical and Experimental Medicine, University of Pisa, Pisa, Italy

M.T. Di Martino
Department of Clinical and Experimental Medicine, Magna Graecia University of Catanzaro, Catanzaro, Italy; Medical Oncology Unit, Mater Domini Hospital, Catanzaro, Italy

M. D'Apolito
Department of Clinical and Experimental Medicine, Magna Graecia University of Catanzaro, Catanzaro, Italy

D. Fanale
Department of Surgical, Oncological, and Oral Sciences, University of Palermo, Palermo, Italy

L. Felicioni
Center of Advanced Studies and Technology - University of Chieti, Chieti, Italy

L. Fiorillo
Medical Oncology Unit, Mater Domini Hospital, Catanzaro, Italy

A. Fiorino
Department of Surgical, Oncological, and Oral Sciences, University of Palermo, Palermo, Italy

A. Galvano
Department of Surgical, Oncological, and Oral Sciences, University of Palermo, Palermo, Italy

A. Giordano
Sbarro Institute for Cancer Research and Molecular Medicine, and Center for Biotechnology, College of Science and Technology, Temple University, Philadelphia, PA, United States

A. Giurintano
Department of Surgical, Oncological, and Oral Sciences, University of Palermo, Palermo, Italy

M. Greco
Department of Surgical, Oncological, and Oral Sciences, University of Palermo, Palermo, Italy

V. Gristina
Department of Surgical, Oncological, and Oral Sciences, University of Palermo, Palermo, Italy

F. Iacono
Department of Surgical, Oncological, and Oral Sciences, University of Palermo, Palermo, Italy

L. Incorvaia
Department of Surgical, Oncological, and Oral Sciences, University of Palermo, Palermo, Italy

M. La Mantia
Department of Surgical, Oncological, and Oral Sciences, University of Palermo, Palermo, Italy

E. Lianidou
Analysis of Circulating Tumor Cells Laboratory, Department of Chemistry, University of Athens, Athens, Greece

U. Malapelle
Department of Public Health, University of Naples Federico II, Naples, Italy

A. Marchetti
Center of Advanced Studies and Technology - University of Chieti, Chieti, Italy

A. Navicella
Center of Advanced Studies and Technology - University of Chieti, Chieti, Italy

E. Pedone
Department of Surgical, Oncological, and Oral Sciences, University of Palermo, Palermo, Italy

A. Perez
Department of Surgical, Oncological, and Oral Sciences, University of Palermo, Palermo, Italy

P. Pisapia
Department of Public Health, University of Naples Federico II, Naples, Italy

A. Pivetti
Department of Surgical, Oncological, and Oral Sciences, University of Palermo, Palermo, Italy

C. Rolfo
Center for Thoracic Oncology, Tisch Cancer Institute, Mount Sinai Medical System & Icahn School of Medicine, Mount Sinai, NY, United States

R. Rossetti
Center of Advanced Studies and Technology - University of Chieti, Chieti, Italy

A. Russo
Department of Surgical, Oncological, and Oral Sciences, University of Palermo, Palermo, Italy; Department of Biomedicine, Neuroscience and Advanced Diagnostics (Bi.N.D.), University of Palermo, Palermo, Italy

R. Scalia
Department of Surgical, Oncological, and Oral Sciences, University of Palermo, Palermo, Italy

V. Spinnato
Department of Surgical, Oncological, and Oral Sciences, University of Palermo, Palermo, Italy

N. Staropoli
Medical Oncology Unit, Mater Domini Hospital, Catanzaro, Italy

P. Tagliaferri
Department of Clinical and Experimental Medicine, Magna Graecia University of Catanzaro, Catanzaro, Italy; Medical Oncology Unit, Mater Domini Hospital, Catanzaro, Italy

P. Tassone
Department of Clinical and Experimental Medicine, Magna Graecia University of Catanzaro, Catanzaro, Italy; Translational Medical Oncology Unit, Mater Domini Hospital, Catanzaro, Italy

S. Taverna
Institute for Biomedical Research and Innovation (IRIB-CNR), National Research Council of Italy, Palermo, Italy

G. Troncone
Department of Public Health, University of Naples Federico II, Naples, Italy

V. Uppolo
Department of Clinical and Experimental Medicine, Magna Graecia University of Catanzaro, Catanzaro, Italy

Preface

Clinical oncology is a rapidly evolving field. During the last several decades, the achievements made empowered continuous improvements in clinical oncology's sphere of influence. Several targeted therapies and immunotherapies are changing the clinical landscape and the natural history of many tumors, impacting patients' survival. In this rapidly evolving scenario, the liquid biopsy of biological fluids (e.g., plasma and serum, urine, saliva, cerebrospinal fluid, pleural fluid, ascites, and stool) is a powerful tool for noninvasive diagnosis, screening, prognosis, and stratification of cancer patients.

Different specialists in the field have covered many aspects of liquid biopsy in this textbook, providing a critical comprehensive overview of this novel field.

Moreover, the main aim of this textbook was to highlight the importance of the cutting-edge liquid biopsy technologies that promise to revolutionize clinical oncology practice. The book covers all basic approaches in the field, explaining their uses, benefits, and limitations.

Notably, the textbook focuses on translational aspects with a deep insight into precision medicine and a comprehensive overview of the arising next-generation sequencing methodologies and associated applications.

Furthermore, even while advances and approvals in oncology are fast, oncologists and all healthcare providers need to be aware, understand, and apply the basic principles and knowledge of liquid biopsy in daily practice.

In this light, we are confident that this book will provide direction to students, oncology residents, and PhD students to think and act accordingly.

Antonio Russo
Ettore Capoluongo
Antonio Giordano
Antonio Galvano

CHAPTER 1

What is precision medicine in oncology?

M. Arbitrio[1,*], A. Cordua[2,*], V. Uppolo[2], M. D'Apolito[2], D. Caracciolo[2], N. Staropoli[3], O. Cuomo[2], L. Fiorillo[3], P. Tassone[2,4], M.T. Di Martino[2,3] and P. Tagliaferri[2,3]

[1]Institute for Research and Biomedical Innovation (IRIB), Italian National Council (CNR), Catanzaro, Italy
[2]Department of Clinical and Experimental Medicine, Magna Graecia University of Catanzaro, Catanzaro, Italy
[3]Medical Oncology Unit, Mater Domini Hospital, Catanzaro, Italy
[4]Translational Medical Oncology Unit, Mater Domini Hospital, Catanzaro, Italy

1.1 Introduction: what does precision medicine mean?

The locution "Precision Medicine" is nowadays very commonly used in several areas of medicine, because of the great burden of its purposes, aims, and political interests. Precision Medicine definition focuses on a deep stratification of patients through the analysis of anthropometrical parameters and biomarkers, both measurable and non-measurable. In particular, it encompasses two phases: (1) data processing and analysis; (2) obtaining and summarizing of the results. In the recent past, this locution was interpreted as "personalized medicine," yet not completely correct, because every medical approach to a patient has to be—by definition—"patient-tailored." Due to these important assumptions, it emerges a clear need to clarify what "precision medicine" exactly means [1]. According to the Presidents' Council of Advisors on Science and Technology "precision medicine" can be defined as "[...] the tailoring of medical treatment to the individual characteristics of each patient. It [...] [means] the ability to classify individuals into subgroups that differ in their susceptibility to a particular disease or their response to a specific treatment" [2]. The aim of precision medicine is to find optimal estimates for prognosis or prediction that could be applied to each individual. It is an approach that leads to obtaining prediction models and treatments which work for the individual patient and who have been obtained through rigorous scientific methods, thus well behind the simple thought of the managing clinician. One main

* These authors contributed equally to the work.

Liquid Biopsy
DOI: https://doi.org/10.1016/B978-0-12-822703-9.00002-8

© 2023 Elsevier Inc.
All rights reserved.

example of such an approach is the use of biomedical markers to better define the baggage of signs and symptoms, in the order to obtain useful information not only for the clinical assessment but also for the definition of an individual therapeutical approach. This point-of-view is nowadays widely used in oncology, because of its largest "medicine cabinet." Owing to these issues, oncology switched from the ideological "one-size-fits-all" theory to a new personalized way of considering a patient as an individual with peculiar characteristics, defined as biomarkers or other clinical parameters.

1.2 The role of biomarkers in precision medicine

The term "personalized" has been accompanied by the definition of a biomarker. According to both the United States National Institute of Health (NIH) and the Food and Drug Administration (FDA), a biomarker indicates a punctual "characteristic that is measured as an indicator of normal biological processes, pathogenic processes, or responses to an exposure or intervention, including therapeutic interventions" (http://www.ncbi. nlm.nih.gov/books/NBK326791). The definition of a biomarker is not only important for the determination of diagnosis or prognosis of a type of cancer: nowadays it is also used to define the pathophysiological main features of cancer. The first example of targeted therapy against the foundation of a single biomarker is defined by imatinib (Gleevec), a tyrosine kinase inhibitor (TKI) of the extracellular domain of Bcr-Abl discovered and introduced in the clinical practice in the 90 teens. Starting from that moment, a lot of molecules were tested in order to find a potential targeted drug against proteins playing a pro-tumor activity [3].

1.3 Classification of biomarkers

There are different kinds of biomarkers, classified according to:
- Structural or functional criteria: genomics (DNA, mtDNA, RNA, mRNA, miRNA, ncRNA), proteomics (proteins, peptides, antibodies), metabolomics (lipids, carbohydrates, enzymes, metabolites);
- Functional clinical target: diagnosis, prognosis, screening, staging, stratification of patients, etc (Fig. 1.1);
- Their clinical role: risk biomarker, predictive, prognostic, surrogate, etc (Fig. 1.2).

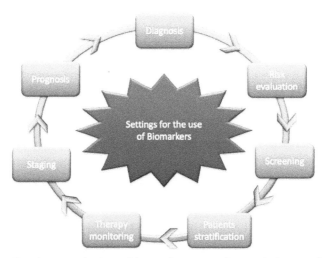

Figure 1.1 Classification of principal biomarkers, according to their own clinical role.

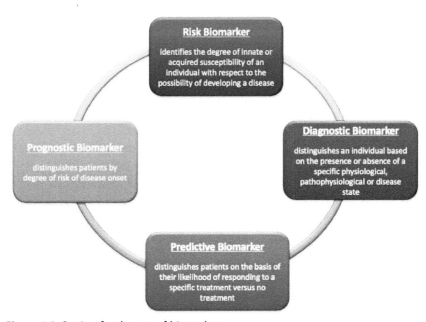

Figure 1.2 Setting for the use of biomarkers.

In particular, a *prognostic biomarker* has to distinguish patients based on their risk of disease onset or of the progression of a specific pathological aspect. For instance, as an example of a prognostic biomarker in common clinical practice, it is possible to indicate the mutation V600E detected in the BRAF gene, which is known to affect the prognosis of patients with melanoma or colorectal cancer. On the other hand, a *predictive biomarker* has the role of distinguishing patients based on the likelihood of response to a particular treatment. For instance, the BRCA genes are universally known as predictive biomarkers: patients with breast cancer, in an adjuvant setting, may benefit from the use of platinum-based chemotherapy regimens; conversely, patients with breast, pancreatic, prostate, and ovarian cancer may benefit from PARP-inhibitors treatment, due to their specific synthetic lethality.

1.4 The rationale for the definition of tumor mutational burden

Nowadays, the most accredited theory on cancer origin is the mutational model, which considers the genetic or somatic mutations on critical genes to be on basis of the cancer development. There are two types of mutation that can be described:

- Driver mutation, which can guarantee a growth advantage of neoplastic cells;
- Passenger mutation does not involve critical genes in cellular growth.

In this context, it is important to define the mutational cancer profile, to detect *actionable* or *druggable mutations*, that can be used as a molecular specific target of molecularly targeted drugs. Moreover, a great deal of interest has emerged today in the definition of the tumor mutation burden (TMB); indeed, it is known that the mutational tumor load is directly proportional to the exposure of neo-antigenic peptides on Major Histocompatibility Complex class I, that make cells recognizable by the immune system and tumors susceptible to the therapeutic use of Immune Checkpoint Inhibitors (ICI) [4]. The TMB aims to define the complexity of the somatic mutations that affect the megabases of specific genomic sequences, thus estimating the mutational tumor load; the latter widely differs across different kinds of neoplasms [5]. Therefore, the role of TMB is to combine scientific research with common clinical practice, thus bridging them [6]. For example, from this point of view, TMB can be interpreted as a predictive response biomarker, capable of choosing patients that can benefit from treatment with ICI, thanks to the achievement of microsatellite instability (MSI), both in colorectal and non-colorectal cancer [7]. Its multidisciplinary approach leads to new therapeutic challenges and perspectives. The duty of TMB has to be considered more important when rarer are the types of mutations on basis of a specific kind of cancer. Since it is an avant-garde approach, it is very important to choose patients that would benefit from these molecular investigations. To solve these practical problems, it is possible to draw up a list of patients, who are eligible for these approaches according to a multidisciplinary group. Although there are age or origins or situs restrictions, patients who have a life expectancy of fewer than six months are excluded from the TMB analysis; patients with other therapeutic chances are excluded, too. In order to make the analysis applicable it is important to collect an adequate sample of biopsy. The traceability of all clinical data must be guaranteed. The increasing interest in studying genome analysis produced new unexpected biomarkers in oncology as in other branches of medicine [8]. For instance, ncRNAs are a class of circulating RNAs under investigation as novel biomarkers for diagnosis or disease monitoring [9—11]. Beyond the emerging interest for ncRNAs as therapeutics targets [12—14], recent evidence indicates that they are expressed in a cell- and tissue-specific pattern, are specifically deregulated in cancer [15], and are released in body fluids in a free form or encapsulated in extracellular vesicles [16]. Moreover, ncRNAs are a heterogeneous class of RNA molecules in terms of chemical structure or biological function, emerging as important mediators in drug sensitivity and drug-resistance mechanisms. Another important frontier was the evaluation of epigenetic modifications, which can be defined as an alteration in DNA methylation or demethylation, that seem to be, their selves, detectable markers of early-onset and progression of cancer [17]. However, despite the initial enthusiasm, it was realized that the prospective use of biomarkers had to be tested by new randomized clinical trials, whose approach needed to be tailored itself to the purpose of producing a new Evidence-Based Clinical Medicine to apply to the use of biomarkers, in order to reclassify tumors [18]. So, it was possible to define new research purposes, concerning new biomarkers-driven approaches, in order to answer many target-related questions [19].

1.5 Collecting samples for mutational analysis: tissue or liquid biopsy?

This novel interest in detecting biomarkers was both driven by the intent of detecting new therapeutical approaches and the definition of early detection of different types of cancer. With these premises, a working group developed a mapping review aimed at the systematic analysis of markers involved in the initial stages of tumor development, thus becoming a milestone in this particular field [20]. Nowadays, in order to make "precision medicine" possible, different ways of sampling are accepted; particularly patients can be subjected to the sampling of a piece of tissue directly by the lesion through a tissue biopsy; more recently a novel approach is based on the sampling of peripheral blood, thus searching, with sensitive techniques, new driving mutations [21]. This modality of sampling is now called liquid biopsy: through this process, a small aliquot of peripheral blood can be sampled in order to research both tumor circulating cells (CTC) and a free double strand of DNA released by tumor circulating cells (ctDNA) or exomes [22]. Although tissue biopsy is widely seen as the gold standard in the definition of the TMB, its invasive approach does not make possible a frequent execution, because of patient discomfort due to the procedure. The finding of ctDNA has the advantage of simple sampling, because of its non-invasive way to approach. It also allows the real-time detection of novel mutations, also response to therapy and surgical effects. However, this approach is limited by the levels of ctDNA which are often low or undetectable [23]. This problem can be solved through the detection of the best time to sample the peripheral blood. The aim of this chapter is to give a "key-of-reading" about the actual connection between liquid biopsy and Precision Oncology.

As already mentioned, liquid biopsy consists of the collection of blood or other biological samples (such as urine, sputum, saliva, pleura fluid, cerebrospinal fluid, etc.), where ctDNA is detectable [24], in order to improve both diagnosis or early detection—even during follow-up—of TMB and biomarkers linked to cancer. Because of the minimum circulating quantity of ctDNA, it is necessary to have sensitive methods, thanks to whom these small and few molecules will become detectable. In fact, ctDNA amounts above the $<0.01\%$ of the total cell-free DNA (cfDNA), especially referring to early stages of cancer [25,26].

Whether it is tissue or liquid biopsy, the pre-processing phase is the most important one. Collecting an adequate sample—both qualitatively & quantitatively—is important to determine the likelihood to detect a mutation that could benefit from targeted therapy. The definition of the accuracy of the sample is a prerogative of the pathologist. In order to collect a useful amount of nucleic material, the number of vital cells has to be greater than the minimum required threshold.

1.6 Pragmatical aspects of precision medicine: how to build it?

According to Literature, the bases for building precision medicine are as complex as the definition already suggests. Indeed, the stratification of patients which is considered the basis of precision medicine is a step-by-step process that moves from the analysis of *deep profiling phenotypes*. In order to be able to proceed with this analysis, the investigated population needs to have features and biomarkers that are common in the general population, the object of the study. So, sampling correctly is very important as the first step of this process. Finding a correct *sample size* allows to give a relevant answer to the research question; the calculation of sample size, as commonly known, has to be done before starting the study [27]. After that, deep phenotyping data need to be processed, in order to create either a diagnostic or a prognostic model. Finally, the last step forecasts, by using the information derived in the previous step, the elaboration of precise therapeutic strategies [1]. Although generally both in observational studies and in clinical trials is necessary to define a population whose results are generalizable to potentially all patients with the disease, in Precision Oncology this is not completely true, because of the intention of evaluating a single subject with all his baggage of biomarkers and molecular driver alterations. Each one of these driver mutations has to be evaluated in order to define a risk stratification approach, whose aim is to define a prediction risk using as starting point a model with or without the new marker of interest. By definition a risk prediction marker is a biomedical parameter both measurable or non-measurable that predicts the punctual risk of a person of developing a determinate event, being—itself—a risk factor. Hence the importance of defining a risk prediction model, that is a single model that provides the information deriving from each marker in order to prelude to the definition of risk; this becomes possible through the stratification of the patient according to these markers. The prediction model has to be tested using three important parameters:

- Calibration is the measurement of the way the calculated risk reflects the fraction of people who had the signaled event;
- The capacity of stratification is the proportion of the population that is correctly stratified in categories with a different risk;
- Classification accuracy, such as the measurement of people who had an event is classified in a high-risk category, while people who had not that event are included in a low-risk category [28].

In order to better define both diagnostic and prognostic models in literature a lot of tools are used; among these, also in observational studies, a central role is played by Receiver Operating Characteristic curves. These curves can define the test capacity to discriminate between diseased or non-diseased individuals. Also, this method has to be subjected to its own calibration to assure the connection with the event [29].

1.7 Precision medicine and clinical trials: is there something different?

As for observational studies, it was clear the traditional clinical trial designs were not able to define the complexity of intervention studies in the period of Precision Oncology, leaving any kind of "reductional approach." Nowadays, more and more oncological patients are subjected to genomic profiling in order to define new several driver mutations that can make them eligible to target therapy, through the methods of Next Generation Sequencing (NGS) platforms [30]. The clinical trials became to be designed on the basis of a biomarker-presence approach (Table 1.1) [19,31]. Among these approaches, the most used is the *enrichment design (adaptive studies)*, which provides the randomization only of patients with the molecular driver alteration selected or on the basis of the response to a specific treatment [32]. Another sub-classification is made between biomarker-stratified and biomarker-based designs. The difference between those two main strategies is made by the fact that in the last approach patients underwent both identifications of a punctual biomarker and, only after that, they were subjected to targeted treatments. On the other hand, the biomarker-stratified approach bypasses the first stage, going straight to randomization and treatment. The adaptive studies represent an evolution of the traditional designs that planned to randomize patients regardless of their molecular status or the presence/absence of a punctual biomarker. This approach is called *randomize-all* [33]. Master protocols are pooled analyses used as experimental platforms of enrichment and randomized-all sub-studies (in this case it is important to add a biomarker negative subgroup of patients), which are similar in key designs and in operational interests [34]. Among these kinds of protocols, it is possible to distinguish:

- *Platform trials*, these studies consist of hierarchical structures where patients with different types of disease are enrolled in a unique study (platform). Next, based on the expression of a biomarker or a response to a specific treatment or a histological type of disease, small groups of

Table 1.1 Principal Features of different biomarker-driven clinical designs.

	Adaptive design	Basket design	Enrichment design	Randomize-all	Umbrella trial
Histology	Dependent	Independent	Dependent	Dependent	Dependent
N. of target therapies	≥ 1	1	1	1	> 1
N. of biomarkers	≥ 1	≥ 1	1	1	> 1
Biomarker type	Biomarker + and biomarker −	Usually biomarker +	Biomarker +	Biomarker + and biomarker −	In case of exploratory trial → biomarker + in case of confirmatory trial → biomarker + and biomarker −
Biomarker a priori credentials	It depends on the study sub-design	Very strong	Very strong	It depends on the study sub-design	Strong
Biomarker assay	Single, locally	Single, locally	Single, locally	Single, locally	Multiplex, centralized
Larger sample size depending on the rarity of driver mutation	+ / −	+ +	+	+ +	+ + +
Statistical complexity	+ + +	+ +	+	+	+ +
Efficiency in time and cost-saving	+	+ +	− −	− −	+ +

similar patients are randomized in treatment vs. placebo study. Interestingly, if the new treatment then shows to be effective, this can be integrated as new best clinical practice. Conversely, if patients do not respond to the novel treatment, they may continue the study (here the name of "eternal trials") and may be included in further randomization trials during the time.

- *Umbrella design trials*, whose aim is to test different types of targeted therapies in different subgroups of the patient (who share the same cancer histological features), are stratified using a biomarker matched approach. This study has to be meticulously planned because of its complexity. A downside of this trial design is the eventual enrollment of a patient with more than a biomarker.
- *Basket design trials*, whose principal feature is being "histology-independent"; in fact, different cancer histological types share the same biomarkers, thus being sensitive to the same type of targeted therapy. Their best challenge is discovering different responses to the same treatment, although sharing the same biomarker. Their downside is the difficulty found in the case of rare cancer types [35].

A milestone was set in 2018 about the changing of indication in using pembrolizumab, a humanized antibody that targets the programmed cell death protein-1 (PD-1) for patients carrying tumors with defects in mismatch repair independently from the site of origin (although it is not yet approved by EMA) [36]. Starting from this epochal change, oncologists became to be interested in looking for single genomic alterations. Hence, the number of intervention studies increased, accelerating every year [19].

1.8 Precision medicine in oncology: what is its "area-of-application"?

The challenge of precision medicine in Oncology started, as already told, with the definition of Bcr-abl as a fusion protein that could be blocked by imatinib, a paradigmatic molecule that functions as a TKI. Since that moment, the interest in finding new biomarkers to use as the molecular target was increasingly higher. Nowadays several lines of cancer research are open and move from the individuation of a biomarker. For instance:

1. Definition of driver molecular alterations in Non-Small Cell Lung Carcinoma (NSCLC), in order to differentiate Oncogenes-addicted NSCLC from Non-Oncogene-addicted NSCLC;

2. Determination of Programmed death-ligand 1 (PD-L1) expression in NSCLC in order to choose patients eligible for immunotherapy;
3. Evaluation of TMB and MSI in colorectal cancer and other types of cancer as biomarkers sensitive to immunotherapy;
4. Definition of synthetic lethality as a target endpoint exploited by using PARP inhibitors, because of the sensitivity of several types of cancer to these drugs.

1.9 Pharmacogenomics and precision medicine

In the optic of precision medicine, Pharmacogenomics (PGx) represents a relevant field of study. It's well known that not all patients respond to the same drug in the same way, and this unpredictable interindividual variability is influenced by several factors including differences in the genetic makeup. Today, specific FDA recommendations for over 200 drugs are provided in their datasheet with mandatory or recommended PGx biomarkers testing before starting therapy (https://www.fda.gov). This variability affects several therapeutic areas (neurological, cardiovascular, analgesic, antiviral, and psychiatric drugs), but cancer therapy has an important impact in terms of dosing, drug efficacy/toxicity, hypersensitivity reactions, and drug resistance and clinical outcome, with a major impact on healthcare costs. To avoid or minimize adverse drug reactions (ADRs), which can impact patient morbidity and mortality, the prescription tailored to the patient instead of a one-size-fits-all approach contributed to precision medicine [37]. The progress in high throughput technologies, including Single Nucleotide Polymorphism (SNP) genotyping and microarray until deep sequencing, highly contributed to the implementation of personalized medicine in clinical practice and the identification of germline and somatic genomic variants [38–40]. However, the identification of a pharmacological phenotype could be of clinical utility. Drug' pharmacokinetics (PK) and pharmacodynamics (PD) can be influenced by polymorphic variants in genes involved in drug absorption, distribution, metabolism, and excretion (ADME) with a strong impact on patient outcomes. Known examples of genomic variants in ADME genes influencing drug PK are $CYP2C19^*17$, associated with bleeding during clopidogrel therapy; $DPYD$ variants, correlated to toxicity risk of 5-fluorouracil or capecitabine; $UGT1A^*28$, associated with irinotecan toxicity; $TPMT$ variants, linked to thiopurine toxicity or among drug transporters the $SLCO1B1^*5$ variant associated with an increased risk for simvastatin toxicity. Instead, related to PD two examples are *HLA-B*5701* associated with the risk of abacavir hypersensitivity or *VKORC1* variants associated with warfarin resistance. The screening of these germ-line gene variants before starting with therapy should allow a reduction of several ADRs or low efficacy of treatment as well as the research of somatic markers in cancer patients for a more tailored prescription and to avoid drug resistances [41]. The agnostic identification of PGx biomarkers, in common and uncommon diseases [42−48] and subsequent external validation need a complex process to overcome the validation and qualifi-cation steps for a biomarker assay development and analytical validation. During a biomarker clinical validation study design the end points, data analysis and reproducibility might be the key source of bias. Prospective larger-scale trials in an independent population or prospective-retrospective study as well as enriched expansion cohorts based on biomarker-adaptive threshold design could allow the achievement of PGx biomarker translation in clinical practice [49]. Moreover. regulatory guidelines (guideline on good pharmacogenomic practice European Medicines Agency (EMA)/Committee for Medicinal Products (CHMP)/ 718998/2016) are available to avoid methodological and statistical issues. However, the implementation of PGx in clinical practice is until now still limited and several barriers (ethical, legal, social) need to be overcome the demonstration of the usefulness and cost-effectiveness of PGx test for clin-icians might allow translation in clinical practice [50].

1.10 Determination of programmed death-ligand 1 expression in non-small cell lung carcinoma in order to choose patients eligible for immunotherapy

Lung cancer nowadays is the most common cause of death all over the world. Almost 85% of the malignancies affecting the lungs are represented by NSCLC, including lung adenocarcinoma (LUAD) and squamous cell lung cancer (LUSC) [51]. From a molecular point of view, this kind of malignancy is heterogeneous: in this perspective, molecular characteriza-tion is crucial in order to guide the correct treatment. In fact, the thera-peutic approach to lung cancer has radically changed in the last twenty years, evolving from cytotoxic therapies to targeted drugs, that act against a singular molecular alteration or the expression status of PD-L1 [52]. These are defined as factors that can also predict the response to targeted drugs or immunotherapy, respectively [53]. Current guidelines specify that every patient affected by NSCLC who is not eligible for loco-regional treatments (for instance, IIIB/IIIC or IV disease stage) has to be subjected to morphological diagnosis with the molecular evaluation of EGFR, BRAF, translocation of ALK or ROS1 and, finally the expression levels of PD-L1. In most cases, these assessments are made on neoplastic tissue samples or cytological samples; however, when tissue sampling is not available or in case of resistant pathology, it is possible to employ ctDNA collected by peripheral blood as a valid alternative [54]. In the pathogenesis of LUAD, the initial oncological step seems to be represented by the KRAS and EGFR mutations, which are blocked by target drugs. KRAS and BRAF are commonly mutually exclusive from each other, but when they coexist, the mutation in KRAS determines resistance to drugs that act against EGFR. On the other hand, in the pathogenesis of LUSC, the most common mutated genes include oncosuppressor, such as TP53 (present in more than 90% of cases) and CDKN2A.

Although even this histotype is characterized by the amplification of EGFR, the activating mutation are rarely seen. An important role in oncological initiation and progression is played by the remodeling of a microenvironment of the malignancy. NSCLC, especially metastatic NSCLC, is characterized by an important tumor mutation load. Some of these mutations lead to the formation of neoantigens, recognized by T-lymphocytes that infiltrate the tumor. The high mutational load results in a higher expression of antigen-presenting molecules and in negative regulators of lymphocytes' activity, like PD-L1 and Lymphocyte-activation gene 3 (LAG-3). Similarly, the presence of MSI is a condition with a high mutational load. These phenotypes can confer sensitivity to immunotherapy.

Surgical resection is the best therapy in stage I, II, and some selected stage IIIA patients. The therapeutic standard for locally advanced disease is surgery and radiotherapy, possibly with a concomitant approach. Platinum-based doublet chemotherapy has long been the standard of care in advanced disease. The advent of target therapy in the late 1990s has indeed transformed the management of advanced lung disease. Today, the therapeutic choice cannot be separated from the evaluation as well as the patient's performance status and tumor histology, the evaluation of the presence of driver mutations, and the level of expression of PDL1. Several NSCLC mutations affect tumor biology and some of them represent a therapeutic target. Among the most frequent are mutations of KRAS (20%−30%), EGFR (10%−15%, but up to 40% of Asian patients), rearrangements of ALK (3%−7%), ROS1 (1%−2%), RET (1%−2%), NTRK (0.5%−1%), BRAF mutations (2%−4%) and HER2 (2%), MET amplifications or mutations (2%−4%) [55]. As a consequence of that, patients with non-squamous or mixed histology, or even squamous if young, non-smokers, should undergo molecular evaluation in order to define the best possible therapy. Mutations in the ATP-binding pocket of EGFR (epidermal growth factor receptor), a member of the family of receptors with tyrosine kinase activity, lead to its constitutive activation with independence from the ligand. In patients with classic mutations (overall representing 90% of EGFR mutations), EGFR inhibitors represent the first line of treatment. These mutations that confer sensitivity to EGFR TKIs are represented by deletions of exon 19 and point mutations of exon 21 (L858R). The remaining 10% is represented by uncommon or rare mutations, including point mutations of exons 18, 19, 20, and 21 or insertions of exon 20.

In stage IV patients with EGFR mutations, TKIs were found to be superior to chemotherapy by improving survival. In particular, first-generation EGFR inhibitors, such as gefitinib and erlotinib, have shown advantages in terms of progression-free survival (PFS) and objective response rates in previously untreated patients when compared with cytotoxic therapy. The second-generation inhibitors, which unlike the first ones are irreversible inhibitors of ATP, showed improvement in PFS and, as in the case of afatinib, better OS compared to chemotherapy in patients with exon 19 deletion but not in those with L858R mutation [56]. After a variable period (approximately 9−13 months), patients with mutated NSCLC-EGFR treated in the first line with a first or second-generation TKI inhibitor undergo progression. In about half of these, this progression is associated with the secondary T790M mutation in EGFR exon 20 which confers resistance to TKIs. It follows, therefore, that in all patients progressing from the first line with EGFR inhibitors, the appearance of this resistance mutation should be evaluated. In these cases, given the high diagnostic accuracy of the analysis carried out on circulating tumor DNA, for the search for the T90M mutation it is possible to consider in the first instance the liquid biopsy reserving the tests on tumor tissue taken through a new biopsy (where technically possible and accepted from the patient), only in case of negativity of the same [57]. Osimertinib, a third-generation inhibitor, has been shown to be very effective in patients with the T790M mutation, representing the second-line treatment of choice in these patients [58]. Additionally, a randomized study comparing osimertinib with first-generation TKI (erlotinib and geftinib) as first-line treatment in patients with advanced EGFR mutated NSCLC showed that this is associated with a significant improvement in PFS. In consideration of these findings, the United States FDA has approved osimertinib for first-line use in EGFR-mutated NSCLC [59]. There are different mechanisms of resistance to TKIs, among these, there is the amplification in HER2 or MET, mutations in BRAF or PIK3CA, as well as the transformation of the histotype into SCLC. It, therefore, follows that, for the definition of the next therapeutic process, a reassessment of the molecular profile of disease progression is important. Another molecular alteration present in up to 3%−7% of patients with LUAD is the rearrangement of the oncogene ALK with EML4 (or other fusion partners present on the short arm of chromosome 2) with the formation of a protein with tyrosine-kinase activity involved in the processes of cell proliferation and survival. The presence of this rearrangement identifies a subgroup of patients with advanced NSCLC with non-squamous histology as candidates for treatment with first-generation (crizotinib) [60], second-generation (alectnib, ceritinib), and a new generation (lorlatinib, brigatinib) of TK ALK inhibitors. Crizotinib, a competitive ATP inhibitor of the tyrosine kinases ALK, MET and ROS1, was the first TKI to prove effective in this setting subgroup. Comparison studies between alectinib and crizotinib in rearranged ALK patients not undergoing prior treatment demonstrated significant improvement in survival in favor of alectinib which, therefore, is now considered the option of choice in the first line of treatment-naïve patients. Without prejudice to the feasibility of strategies beyond progression and local treatment in case of oligoprogression, as evidenced for the EGFR TKI also for patients undergoing treatment with crizotinib, after a period of approximately 12 months, a clinical-radiological progression occurs attributable to the acquisition of drug resistance. There are many mechanisms involved in drug resistance, including point mutations in ALK, up-regulation of oncogenes, or activation of other intracellular signaling pathways. Also for second-generation inhibitors, disease progression is associated with the development of resistance mechanisms, especially ALK-dependent 318. In the setting of patients progressing from second-generation inhibitors, treatment with lorlatinib appears to be promising, especially due to its marked activity. in the brain, a frequent site of metastases in these patients [61]. In patients who have run out of molecularly targeted treatments, a subgroup analysis of the results of the Empower 150 study led the EMA to approve the combination of carboplatin, paclitaxel, bevacizumab, atezolizumab [62].

Structurally similar to ALK, ROS-1, receptor tyrosine kinase is an oncogene driver in 1%—2% of NSCLCs. Currently, the only drug recognized in patients with rearranged ROS-1 disease is crizotinib. Other inhibitors under study are ceritinib, lorlatinib, entrectinib, and repotrectinib. Lorlatinib and repotrectinib exhibit activity after progression from crizotinib. Entrectnib, on the other hand, exhibits activity in patients with ROS1 and NTRK fusion [63].

V600 point mutations in exon 15 of BRAF, present in approximately 3%—4% of LUADs, represent another predictive biomarker of response to combined treatment with the TKIs dabrafenib and trametinib. This combination should therefore be considered a first- or second-line treatment option in patients with advanced NSCLC with a BRAF V600 mutation [64].

There are other molecular alterations of NSCLC such as MET mutations, KRAS exon 2, HER-2, NTRK1-3, and RET rearrangements. Larotrectinib has been shown to be a treatment option in patients with NTRK gene fusion and can be used in lung cancer patients with this gene fusion [65]. Activating mutations in the tyrosine kinase domain of HER2 appear to confer sensitivity to afatinib therapy. FGFR1 amplifications, PI3KCA mutations, PTEN, DDR2, and PDGFR amplifications may play a role in the future.

In patients without target therapy-susceptible driver mutations, platinum-based first-line chemotherapy was the standard of care. The advent of immunotherapy has radically changed the treatment of the disease in this patient setting. For this reason, the PDL-1 expression level is one of the predictive markers to be tested in patients with advanced NSCLC. The purpose of this evaluation is to identify patients eligible for first-line immunotherapy treatment with pembrolizumab. Pembrolizumab is the first choice in the treatment of patients with PDL1 expression greater than 50% of tumor cells evaluated according to the Tumor Proportional Score on at least 100 tumor cells in formalin-fixed and incused tissue, cytological and histological specimens. in paraffin 60. For access to the second therapeutic line, the level of PD-L1 expression required is instead greater than 1%.

1.11 Through the concept of synthetic lethality: poly ADP-ribose polymerases-inhibitors and precision medicine

Poly ADP-ribose polymerases (PARP poly) are a superfamily of proteins involved in several cellular processes such as stress response, chromatin remodeling, DNA repair, and apoptosis [66]. Owing to the multiple molecular pathways in which PARP is involved, it has been considered an important pharmacological target in oncology. In particular, PARP1 is a gene, mapping on chromosome 1q41.42, that regulates cell proliferation and differentiation. Hence, it is crucial during single-strand break repair, and damaged purine bases. It is also known that the product protein of this gene has an important affinity for double-strand breaks. Together with tP53, KRAS, Myc, and other genes, PARP leads to the reparation of single or double-strand breaks on nucleic acid [67]. Despite its important role, mice models knocked out for PARP1 have a delayed tumor presentation; this phenomenon is probably attributable to PARP2 protein (another member of the PARP superfamily), which substitutes the role of PARP1, because of their overlapping functions [68]. *Synthetic lethality* is the simultaneous occurrence of mutations on different genes, that finally result in cell death [69]. Another important role is played by homologous recombination repair defective cells, because of the sequestration of PARP on DNA, and, subsequently, cellular sensitization [70]. Since 1946, when Theodosius Dobzhansky first developed the concept of synthetic lethality, a great interest increased in the role of PARP inhibitors. The first phase I trial involving the use of olaparib, an oral active molecule that acts as a PARP inhibitor, was published in 2009 in the New England Journal of Clinical Medicine [71]. Nowadays different types of PARP inhibitors are approved for clinical use not only for ovarian cancer but also in patients with breast cancer [72,73]. This clinical trial pointed out the reason why patients who are carriers of BRCA1 or BRCA2 germline mutations. In fact, these patients have a higher risk of developing both breast and ovarian cancer if female and breast and prostate cancer if male [74]. From a biological point of view, cells heterozygous for BRCA mutation, can also lose their remaining wild type allele, thus producing a deficiency in homologous-recombination repair. The consequent homologous recombination deficiency (HRD) is harbored by approximately 13% and 15% of ovarian and triple-negative breast cancer (TNBC) respectively, due to the germline mutations in BRCA1/2. Among these cases, 50% and 40% of ovarian and TNBC are characterized by harboring HRD in the absence of BRCA1/2 [75]. Interestingly, the use of PARP inhibitors as maintenance therapy in a population of women with relapsed high-grade ovarian cancer determined an enhancement of response to platinum-based chemotherapy regimens, as it was assessed by several phase 2 or 3 clinical trials [76]. In this biological dimension, PARP inhibitors have the precious role of determining the accumulation of DNA single-strand breaks, which determine, in turn, double-strand breaks and, in the end, the collapse of replication forks [77]. Nowadays, apart from olaparib, the oral active molecule which led the way to the "PARPihera," also veliparib, recuparib, talazoparib, and pamiparib, thanks to their registration studies, are approved for a specific indication and posology, by the principal drug agencies [78].

DNA repair dysregulation as a cancer driver can be seen as a therapeutic target and biomarker of immunotherapy sensitivity. In fact, genetic alterations in cancer cells could result in the formation of an important load of neoantigens, or genomic aberrations that could also result in increased immunogenicity (i.e. missense mutations, frameshift mutations, stop codon mutations, fusion transcript) [79,80].

International or national guidelines for breast cancer define the role of BRCA germline mutation research in two specific settings: patients in early-stage with additional criteria or diagnosed with metastatic HER2-negative disease, in order to find a therapeutic implication. There are two ways of determining the mutational status:

1. *Single-gene testing*, whose role is finding a mutation-specific gene, using polymerase chain reaction (PCR) or direct sequencing (it was the method traditionally used);
2. *Panel testing*, whose advantage is the simultaneous evaluation of multiple panels of genes, using just one sample through the application of NGS.

1.12 Colorectal cancer e microsatellite instability

The accumulation of genetic alterations such as nucleotide alterations, epigenetic changes, structural changes, and histone modifications are the basis for the transformation of normal cells into cancer cells, therefore, accurate DNA repair is essential to maintain genetic stability [81−83]. DNA repair systems, a fundamental role is played by the mismatch repair system (MMR). Tumors with a defective DNA mismatch repair (dMMR) system are often hyper-mutated and accumulate mutations in short tandem repeats (monomorphic microsatellites).

This condition, termed MSI, represents the phenotypic and molecular evidence of a non-normally functioning MMR that is accompanied by a 100−1000-fold increase in the mutation rate [84,85]. MSI (MSI-H or MSI high) is a molecular feature shared by different types of cancers, especially gynecological and gastrointestinal ones, and guides the pathogenesis of about 15% of colorectal cancers and 5% of those in the metastatic stage; in 3% of cases, it is associated with Lynch Syndrome, while the remaining cases are sporadic [86−90].

MSI is a sign of sporadic or hereditary dysfunction of the MMR pathway determined by various factors, such as mutations of related MMR genes, and inactivation of the gene transcription due to hypermethylation of the promoter region, or repression of transcription caused by inflammation. MSI-positive (MSI +) tumors, regardless of the site of origin, possess molecular, pathological, and clinical characteristics that are different from MSI-negative (MSI −) ones. High MSI colorectal carcinomas are more frequently diagnosed in the right colon and at a young age, at an earlier stage than the stable forms, with the histological-molecular prevalence of the mucinous histotype and poor differentiation, in addition to the mutation of the BRAF gene.

The evaluation of dMMR can be tested using immunohistochemistry (IHC) which allows for diagnosing the lack of expression of the proteins encoded by one of the 4 main genes of the MMR (MLH1, MSH2, MSH6, PMS2) and molecular tests to diagnose MSI (MSIH or MSI high) by PCR and new NGS approaches.

The DNA MMR Repair System is capable of restoring DNA integrity after the occurrence of mismatch errors, including single-base mismatches or brief insertions and deletions. The genes that play a crucial role in this process are MLH1 (mutL homolog 1), MSH2 (mutS homolog 2), MSH6 (mutS homolog 6), and PMS2 (postmeiotic segregation increased 2) encoding four homonymous proteins. Germinal and/or somatic mutations or epigenetic silencing of one of these genes can result in a dMMR mechanism [91]. MLH1 and MSH2 are mandatory partners of heterodimers. PMS2 can form a heterodimer only with MLH1, and MSH6 can form a heterodimer only with MSH2. However, MLH1 and MHS2 can form heterodimers with other MMR proteins, namely MSH3, MLH3, and PMS1. In general, mutations in MLH1 and MSH2 result in the subsequent proteolytic degradation of the mutated protein and its secondary partner, PMS2 and MSH6, respectively. Conversely, mutations in PMS2 or MSH6 may not result in proteolytic degradation of its primary partner, since MSH6 can be substituted in the heterodimer by MSH3 and PMS2 can be substituted in the heterodimer by PMS1 or MLH3.

The presence of MSI has prognostic significance as it has been shown to be associated with better prognosis in patients with resectable colorectal cancer in the early stages of the disease modified only in part by the status of BRAF while in the metastatic stage this advantage seems to be lost in association with a likely intrinsic chemo-resistance [92–101].

The determination of MSI has a supportive role in the decision-making choice to administer adjuvant therapy given the extremely favorable prognosis and a different effect of adjuvant therapy (data still controversial on the possible resistance to 5-Fluorouracil therapy) [102–104].

Several lines of evidence show that MMR/MSI is a predictive marker of response for immunotherapy for CRC but also for other related S. of Lynch tumors [7,105–107]. The most recent evidence has revealed that tumors with MSI-H or dMMR, regardless of their primary site, have a promising response to ICI. The functional defect affecting the MMR causes the accumulation of genetic mutations during DNA replication, thus increasing the mutational TMB and neo-antigenic load, from which an enhanced endogenous immune response and a correlated sensitivity to immunotherapy derive [108–111]. Cancer Genome Atlas analyzes divided CRCs into hyper-mutated and non-hyper-mutated types [112,113]. Of those hyper-mutated (16% of colorectal cancers), one-quarter have mutations in the proofreading (exonuclease) subunit of the epsilon polymerase and three-quarters show MSI-H, usually with methylation of the MLH1 promoter. The Colorectal Cancer Subtyping Consortium classified CRCs into four molecular consensus subtypes (CMS) with distinctive biological characteristics: CMS1 (immunity from MSI, 14%), CMS2 (canonical, 37%), CMS3 (metabolic, 13%), CMS4 (mesenchymal, 23%) and mixed characteristics (13%) [114]. Among them, CMS1 tumors exhibit MSIH features with MLH1 promoter methylation or mutations in MMR-related genes. These tumors show an increase in tumor-infiltrating lymphocytes, mainly Th1 and CTL, and an elevated expression of PD-L1, together with strong activation of the immune evasion pathways. These tumors are potential targets for anti-PD-1 therapy.

It is well established that ICI has revolutionized the treatment of patients with advanced cancer. Based on the results of 149 patients with MSI-H or dMMR enrolled in 5 clinical trials (KEYNOTE-016, KEYNOTE-164, KEYNOTE-012, KEYNOTE-028, and KEYNOTE-158), in May 2017, the FDA granted the accelerated approval of pembrolizumab, an anti-PD-1 monoclonal antibody, for the treatment of adult and pediatric patients with solid, unresectable or metastatic tumors with MSI-H or progressing dMMR from conventional treatments.

This was the first FDA approval based on a biomarker rather than the anatomical location in the body where the tumor originated. Shortly thereafter, nivolumab (Opdivo), an anti-PD-1 monoclonal antibody, also gained accelerated approval in August 2017 for adult and pediatric patients with MSI-H or metastatic dMMR of progressing colorectal cancer after standard chemotherapy.

More recently, evidence has been obtained of the efficacy of immune ICI even in the first-line setting of CRC (CheckMate-142, Keynote 177) [115−117].

Given the important role it can play in particular in relation to the application of immunotherapy for the treatment of cancer, there is a need for versatile and reliable diagnostic procedures for MSI. In recent years, alongside the evaluation based on IHC and on the PCR, the possibility of an analysis based on the use of NGS has been reported [118].

IHC is a widely available and less expensive method for MSI analysis based on the use of antibodies against the four MMR proteins. IHC can be performed on surgical specimens or biopsies. It has been shown that most of the mutations of the MMR genes interfere with the dimerization of proteins with consequent loss of the same due to the proteolytic degradation of the heterodimers.

In the presence of mutations in MLH1, there is a loss of IHC of both MLH1 and PMS2, while for mutations in MSH2 there is a loss of IHC of both MSH2 and MSH6. When mutations occur in secondary protein genes (e.g., PMS2 and MSH6), heterodimers can remain stable and there is no simultaneous loss of binding partner proteins because the function of secondary proteins can be compensated for by other proteins, such as MSH3 instead of MSH6, eMLH3 or PMS1 instead of PMS2. It follows that while the PMS2 antibody detects all cases harboring MLH1 or PMS2 abnormalities and the MSH6 antibody detects all cases harboring an MSH2 or MSH6 abnormality, MLH1 and MSH2 alone do not recognize cases that have PMS2 abnormalities or MSH6. The IHC can give rise to results that do not reflect the real state of MMR for pre-analytical problems such as tissue fixation or for biological reasons related to the type of mutation associated with the dMMR [84].

It is strongly recommended to use all four antibodies, simultaneously or in sequence, and to switch to MSI-PCR whenever there is any doubt in the interpretation of the IHC [84].

Another standard method for diagnosing MSI is the genotyping of microsatellites using PCR-based tests. This method is indicated in cases where there are indeterminate IHC results, interpretation difficulties, or loss of a single heterodimeric subunit (e.g., only MLH1 or only PMS2 and not both).

The molecular test is based on the assumption that MMR deficiency causes the expansion or contraction of microsatellite regions that are used as "microsatellite markers." PCR amplification of microsatellite markers can be performed with two possible panels: the "Bethesda panel" uses five microsatellite markers including two mononucleotide repeats (BAT-25 and BAT-26) and three dinucleotide repeats (D2S123, D5S346, and D17S250), the other panel uses five poly-A mononucleotide repeats (BAT-25, BAT-26, NR-21, NR-24, NR-27) and is considered the current standard due to its higher specificity and sensitivity [119].

The NGS represents an alternative molecular test to evaluate the MSI that has been shown to agree with the 5.8%–100% with PCR-based tests [111–121]. Timmermann et al. demonstrated that deep sequencing of cancer exons can allow prediction of MSI in CRC and detection of clinically significant mutations [122]. "mSINGS" developed by Salipante et al. (method for determining MSI status using NGS) also proved to be an accurate method with high sensitivity and specificity [123]. An important advantage of the NGS approach can be represented by the possibility of evaluating other gene alterations capable of directing therapeutic choices at the same time as MSI. In this direction, Hempelmann et al. developed MSIplus capable of identifying the MSI status of CRC and cancer-related mutations (KRAS, NRAS, and BRAF) [124]. A further advantage over other methods is the contextual evaluation of MSI and TMB, the total number of mutations per coding area of a tumor genome. Campbell et al. showed that hypermutation (defined as >10 somatic mutations per megabase) was more prevalent than previously estimated, affecting approximately 17% of adult cancers. These findings potentially expand the use of immunotherapy, which is believed to be effective in cancers with an increased mutational burden [125].

1.13 Conclusions

The continuous improvement in technology, oncology trials design, and computational methods such as TMB allowed the enhancement of personalized medicine toward precision medicine. As above described the aim of precision medicine is the deep knowledge of all uncovered molecular characteristics of cancer to prescribe a patient's fit therapy based on its specific features for better-managed care. The identification of potential or recognized cancer biomarkers improving outcomes could allow cohort selection and enrollment in precision oncology trials design with the aim to improve and guide the better-tailored prescription and detection of enhanced treatment efficacy in a patient subpopulation. This is the case of patient selection according to a known biomarker or the identification of a driver mutation in gene harbor cancer type-specific. Liquid biopsy and CTCs could represent important support in this context as non-invasive methods and will be of great help to clinical screening, earlier cancer detection and patient stratification to the most suitable treatment, the evaluation of treatment response or tumor resistance, the consideration of progression disease and its risk prognosis related. Diagnosis and tumors screening acquire insights into precision therapy. At the present, the percentage of clinical trials requiring a genetic alteration for enrollment has increased dramatically in Progression Free Survival and Overall Survival improvements. However, although precision medicine is not used for every cancer type, all stakeholders' efforts are needed to allow, one day, the customization of treatments to the specific gene alterations in each specific cancer type allowing the overcoming of actual limitations, both in labs and in clinical trials.

Conflict of interest statement

The authors declare that the research was conducted in the absence of any commercial or financial relationships that could be construed as a potential conflict of interest

References

[1] Konig IR, Fuchs O, Hansen G, von Mutius E, Kopp MV. What is precision medicine? Eur Respir J 2017;50(4).
[2] Toward Precision Medicine: Building a Knowledge Network for Biomedical Research and a New Taxonomy of Disease. The National Academies Collection: Reports funded by National Institutes of Health. Washington, DC; 2011.
[3] Mauro MJ, O'Dwyer ME, Druker BJ. ST1571, a tyrosine kinase inhibitor for the treatment of chronic myelogenous leukemia: validating the promise of molecularly targeted therapy. Cancer Chemother Pharmacol 2001;48(Suppl. 1):S77−8.
[4] Barroso-Sousa R, Jain E, Cohen O, Kim D, Buendia-Buendia J, Winer E, et al. Prevalence and mutational determinants of high tumor mutation burden in breast cancer. Ann Oncol 2020;31(3):387−94.
[5] Sha D, Jin Z, Budczies J, Kluck K, Stenzinger A, Sinicrope FA. Tumor mutational burden as a predictive biomarker in solid tumors. Cancer Discov 2020;10 (12):1808−25.
[6] van der Velden DL, van Herpen CML, van Laarhoven HWM, Smit EF, Groen HJM, Willems SM, et al. Molecular tumor boards: current practice and future needs. Ann Oncol 2017;28(12):3070−5.

[7] Marabelle A, Le DT, Ascierto PA, Di Giacomo AM, De Jesus-Acosta A, Delord JP, et al. Efficacy of pembrolizumab in patients with noncolorectal high microsatellite instability/mismatch repair-deficient cancer: results from the phase II KEYNOTE-158 study. J Clin Oncol 2020;38(1):1−10.

[8] Werner RJ, Kelly AD, Issa JJ. Epigenetics and precision oncology. Cancer J 2017;23 (5):262−9.

[9] Di Mauro S, Scamporrino A, Petta S, Urbano F, Filippello A, Ragusa M, et al. Serum coding and non-coding RNAs as biomarkers of NAFLD and fibrosis severity. Liver Int 2019;39(9):1742−54.

[10] Barbagallo C, Di Martino MT, Grasso M, Salluzzo MG, Scionti F, Cosentino FII, et al. Uncharacterized RNAs in plasma of Alzheimer's patients are associated with cognitive impairment and show a potential diagnostic power. Int J Mol Sci 2020;21(20).

[11] Di Mauro S, Scamporrino A, Fruciano M, Filippello A, Fagone E, Gili E, et al. Circulating coding and long non-coding RNAs as potential biomarkers of idiopathic pulmonary fibrosis. Int J Mol Sci 2020;21(22).

[12] Rossi M, Di Martino MT, Morelli E, Leotta M, Rizzo A, Grimaldi A, et al. Molecular targets for the treatment of multiple myeloma. Curr Cancer Drug Targets 2012;12(7):757−67.

[13] Di Martino MT, Riillo C, Scionti F, Grillone K, Polera N, Caracciolo D, et al. miRNAs and lncRNAs as novel therapeutic targets to improve cancer immunotherapy. Cancers (Basel) 2021;13(7).

[14] Caracciolo D, Montesano M, Altomare E, Scionti F, Di Martino MT, Tagliaferri P, et al. The potential role of miRNAs in multiple myeloma therapy. Expert Rev Hematol 2018;11(10):793−803.

[15] Grillone K, Riillo C, Scionti F, Rocca R, Tradigo G, Guzzi PH, et al. Non-coding RNAs in cancer: platforms and strategies for investigating the genomic "dark matter.". J Exp Clin Cancer Res 2020;39(1):117.

[16] Russo M, Tirinato L, Scionti F, Coluccio ML, Perozziello G, Riillo C, et al. Raman spectroscopic stratification of multiple myeloma patients based on exosome profiling. ACS Omega 2020;5(47):30436−43.

[17] Fojo T. Novel_target.com. Oncologist 2001;6(4):313−14.

[18] Prasad V, Fojo T, Brada M. Precision oncology: origins, optimism, and potential. Lancet Oncol 2016;17(2):e81−6.

[19] Janiaud P, Serghiou S, Ioannidis JPA. New clinical trial designs in the era of precision medicine: an overview of definitions, strengths, weaknesses, and current use in oncology. Cancer Treat Rev 2019;73:20−30.

[20] Uttley L, Whiteman BL, Woods HB, Harnan S, Philips ST, Cree IA, et al. Building the evidence base of blood-based biomarkers for early detection of cancer: a rapid systematic mapping review. EBioMedicine 2016;10:164−73.

[21] Chan KC, Jiang P, Zheng YW, Liao GJ, Sun H, Wong J, et al. Cancer genome scanning in plasma: detection of tumor-associated copy number aberrations, single-nucleotide variants, and tumoral heterogeneity by massively parallel sequencing. Clin Chem 2013;59(1):211−24.

[22] Crowley E, Di Nicolantonio F, Loupakis F, Bardelli A. Liquid biopsy: monitoring cancer-genetics in the blood. Nat Rev Clin Oncol 2013;10(8):472−84.

[23] Wang J, Chang S, Li G, Sun Y. Application of liquid biopsy in precision medicine: opportunities and challenges. Front Med 2017;11(4):522−7.

[24] Peng M, Chen C, Hulbert A, Brock MV, Yu F. Non-blood circulating tumor DNA detection in cancer. Oncotarget 2017;8(40):69162−73.

[25] Norton SE, Luna KK, Lechner JM, Qin J, Fernando MR. A new blood collection device minimizes cellular DNA release during sample storage and shipping when compared to a standard device. J Clin Lab Anal 2013;27(4):305−11.

[26] Franczak C, Filhine-Tresarrieu P, Gilson P, Merlin JL, Au L, Harle A. Technical considerations for circulating tumor DNA detection in oncology. Expert Rev Mol Diagn 2019;19(2):121−35.

[27] Noordzij M, Dekker FW, Zoccali C, Jager KJ. Sample size calculations. Nephron Clin Pract 2011;118(4):c319−23.

[28] Janes H, Pepe MS, Gu W. Assessing the value of risk predictions by using risk stratification tables. Ann Intern Med 2008;149(10):751−60.

[29] Cook NR. Statistical evaluation of prognostic vs diagnostic models: beyond the ROC curve. Clin Chem 2008;54(1):17−23.

[30] Takebe N, Yap TA. Precision medicine in oncology. Curr Probl Cancer 2017;41 (3):163−5.

[31] Renfro LA, An MW, Mandrekar SJ. Precision oncology: a new era of cancer clinical trials. Cancer Lett 2017;387:121−6.

[32] Renfro LA, Mallick H, An MW, Sargent DJ, Mandrekar SJ. Clinical trial designs incorporating predictive biomarkers. Cancer Treat Rev 2016;43:74−82.

[33] Tajik P, Zwinderman AH, Mol BW, Bossuyt PM. Trial designs for personalizing cancer care: a systematic review and classification. Clin Cancer Res 2013;19 (17):4578−88.

[34] Woodcock J, LaVange LM. Master protocols to study multiple therapies, multiple diseases, or both. N Engl J Med 2017;377(1):62−70.

[35] Menis J, Hasan B, Besse B. New clinical research strategies in thoracic oncology: clinical trial design, adaptive, basket and umbrella trials, new end-points and new evaluations of response. Eur Respir Rev 2014;23(133):367−78.

[36] Heymach J, Krilov L, Alberg A, Baxter N, Chang SM, Corcoran RB, et al. Clinical cancer advances 2018: annual report on progress against cancer from the american society of clinical oncology. J Clin Oncol 2018;36(10):1020−44.

[37] Evans WE, Relling MV. Moving towards individualized medicine with pharmacogenomics. Nature. 2004;429(6990):464−8.

[38] Arbitrio M, Scionti F, Di Martino MT, Caracciolo D, Pensabene L, Tassone P, et al. Pharmacogenomics biomarker discovery and validation for translation in clinical practice. Clin Transl Sci 2021;14(1):113−19.

[39] Gallo Cantafio ME, Grillone K, Caracciolo D, Scionti F, Arbitrio M, Barbieri V, et al. From single level analysis to multi-omics integrative approaches: a powerful strategy towards the precision oncology. High Throughput 2018;7(4).

[40] Scionti F, Agapito G, Caracciolo D, Riillo C, Grillone K, Cannataro M, et al. Risk Alleles for Multiple Myeloma Susceptibility in ADME Genes. Cells 2022;11(2) 10.3390/cells11020189.

[41] Arbitrio M, Di Martino MT, Scionti F, Agapito G, Guzzi PH, Cannataro M, et al. DMET (drug metabolism enzymes and transporters): a pharmacogenomic platform for precision medicine. Oncotarget 2016;7(33):54028−50.

[42] Arbitrio M, Di Martino MT, Scionti F, Barbieri V, Pensabene L, Tagliaferri P. Pharmacogenomic profiling of ADME gene variants: current challenges and validation perspectives. High Throughput 2018;7(4).

[43] Di Martino MT, Arbitrio M, Leone E, Guzzi PH, Rotundo MS, Ciliberto D, et al. Single nucleotide polymorphisms of ABCC5 and ABCG1 transporter genes correlate to irinotecan-associated gastrointestinal toxicity in colorectal cancer patients: a DMET microarray profiling study. Cancer Biol Ther 2011;12(9):780−7.

[44] Di Martino MT, Arbitrio M, Guzzi PH, Leone E, Baudi F, Piro E, et al. A peroxisome proliferator-activated receptor gamma (PPARG) polymorphism is associated with zoledronic acid-related osteonecrosis of the jaw in multiple myeloma patients: analysis by DMET microarray profiling. Br J Haematol 2011;154 (4):529−33.

[45] Arbitrio M, Di Martino MT, Barbieri V, Agapito G, Guzzi PH, Botta C, et al. Identification of polymorphic variants associated with erlotinib-related skin toxicity in advanced non-small cell lung cancer patients by DMET microarray analysis. Cancer Chemother Pharmacol 2016;77(1):205−9.

[46] Arbitrio M, Scionti F, Altomare E, Di Martino MT, Agapito G, Galeano T, et al. Polymorphic variants in NR1I3 and UGT2B7 predict taxane neurotoxicity and have prognostic relevance in patients with breast cancer: a case-control study. Clin Pharmacol Ther 2019;106(2):422−31.

[47] Di Martino MT, Scionti F, Sestito S, Nicoletti A, Arbitrio M, Hiram Guzzi P, et al. Genetic variants associated with gastrointestinal symptoms in Fabry disease. Oncotarget 2016;7(52):85895−904.

[48] Scionti F, Di Martino MT, Sestito S, Nicoletti A, Falvo F, Roppa K, et al. Genetic variants associated with Fabry disease progression despite enzyme replacement therapy. Oncotarget 2017;8(64):107558−64.

[49] Dobbin KK, Cesano A, Alvarez J, Hawtin R, Janetzki S, Kirsch I, et al. Validation of biomarkers to predict response to immunotherapy in cancer: volume II—clinical validation and regulatory considerations. J Immunother Cancer 2016;(4):77.

[50] Scionti F, Pensabene L, Di Martino MT, Arbitrio M, Tagliaferri P. Ethical Perspectives on Pharmacogenomic Profiling. In: Ramalakshmi Boobalan, Nicholls Michael, editors. Reference Module in Biomedical Sciences 10.1016/B978-0-12-820472-6.00139-0. Elsevier; 2021. In press.

[51] Socinski MA, Obasaju C, Gandara D, Hirsch FR, Bonomi P, Bunn P, et al. Clinicopathologic features of advanced squamous nSCLC. J Thorac Oncol 2016;11 (9):1411−22.

[52] Osmani L, Askin F, Gabrielson E, Li QK. Current WHO guidelines and the critical role of immunohistochemical markers in the subclassification of non-small cell lung carcinoma (NSCLC): Moving from targeted therapy to immunotherapy. Semin Cancer Biol 2018;52(Pt 1):103−9.

[53] Proto C, Ferrara R, Signorelli D, Lo Russo G, Galli G, Imbimbo M, et al. Choosing wisely first line immunotherapy in non-small cell lung cancer (NSCLC): what to add and what to leave out. Cancer Treat Rev 2019;75:39−51.

[54] Imyanitov EN, Iyevleva AG, Levchenko EV. Molecular testing and targeted therapy for non-small cell lung cancer: current status and perspectives. Crit Rev Oncol Hematol 2021;157:103194.

[55] Lindeman NI, Cagle PT, Aisner DL, Arcila ME, Beasley MB, Bernicker EH, et al. Updated molecular testing guideline for the selection of lung cancer patients for treatment with targeted tyrosine kinase inhibitors: guideline from the College of American Pathologists, the International Association for the Study of Lung Cancer, and the Association for Molecular Pathology. Arch Pathol Lab Med 2018;142 (3):321−46.

[56] Sarosi V, Baliko Z. [Efficacy of first-line afatinib vs chemotherapy in EGFR mutation positive pulmonary adenocarcinoma]. Magy Onkol 2014;58(4):325−9.

[57] Del Re M, Crucitta S, Gianfilippo G, Passaro A, Petrini I, Restante G, et al. Understanding the mechanisms of resistance in EGFR-positive NSCLC: from tissue to liquid biopsy to guide treatment strategy. Int J Mol Sci 2019;20(16).

[58] Leighl NB, Karaseva N, Nakagawa K, Cho BC, Gray JE, Hovey T, et al. Patient-reported outcomes from FLAURA: osimertinib vs erlotinib or gefitinib in patients with EGFR-mutated advanced non-small-cell lung cancer. Eur J Cancer 2020;125:49−57.

[59] Ramalingam SS, Yang JC, Lee CK, Kurata T, Kim DW, John T, et al. Osimertinib as first-line treatment of EGFR mutation-positive advanced non-small-cell lung cancer. J Clin Oncol 2018;36(9):841−9.

[60] Arbour KC, Riely GJ. Systemic therapy for locally advanced and metastatic non-small cell lung cancer: a review. JAMA 2019;322(8):764—74.

[61] Shaw AT, Bauer TM, de Marinis F, Felip E, Goto Y, Liu G, et al. First-line lorlatinib or crizotinib in advanced ALK-positive lung cancer. N Engl J Med 2020;383(21):2018—29.

[62] Manzo A, Carillio G, Montanino A, Sforza V, Palumbo G, Esposito G, et al. The safety of atezolizumab plus chemotherapy for the treatment of metastatic lung cancer. Expert Opin Drug Saf 2020;19(7):775—83.

[63] Drilon A, Siena S, Dziadziuszko R, Barlesi F, Krebs MG, Shaw AT, et al. Entrectinib in ROS1 fusion-positive non-small-cell lung cancer: integrated analysis of three phase 1—2 trials. Lancet Oncol 2020;21(2):261—70.

[64] Planchard D, Smit EF, Groen HJM, Mazieres J, Besse B, Helland A, et al. Dabrafenib plus trametinib in patients with previously untreated BRAF(V600E)-mutant metastatic non-small-cell lung cancer: an open-label, phase 2 trial. Lancet Oncol 2017;18(10):1307—16.

[65] Haratake N, Seto T. NTRK fusion-positive non-small-cell lung cancer: the diagnosis and targeted therapy. Clin Lung Cancer 2021;22(1):1—5.

[66] Strom CE, Helleday T. Strategies for the use of poly(adenosine diphosphate ribose) polymerase (PARP) inhibitors in cancer therapy. Biomolecules 2012;2(4):635—49.

[67] Ashworth A, Lord CJ. Synthetic lethal therapies for cancer: what's next after PARP inhibitors? Nat Rev Clin Oncol 2018;15(9):564—76.

[68] Wang ZQ, Auer B, Stingl L, Berghammer H, Haidacher D, Schweiger M, et al. Mice lacking ADPRT and poly(ADP-ribosyl)ation develop normally but are susceptible to skin disease. Genes Dev 1995;9(5):509—20.

[69] O'Neil NJ, Bailey ML, Hieter P. Synthetic lethality and cancer. Nat Rev Genet 2017;18(10):613—23.

[70] Helleday T. The underlying mechanism for the PARP and BRCA synthetic lethality: clearing up the misunderstandings. Mol Oncol 2011;5(4):387—93.

[71] Fong PC, Boss DS, Yap TA, Tutt A, Wu P, Mergui-Roelvink M, et al. Inhibition of poly(ADP-ribose) polymerase in tumors from BRCA mutation carriers. N Engl J Med 2009;361(2):123—34.

[72] Litton JK, Rugo HS, Ettl J, Hurvitz SA, Goncalves A, Lee KH, et al. Talazoparib in patients with advanced breast cancer and a germline BRCA mutation. N Engl J Med 2018;379(8):753—63.

[73] Dieras V, Han HS, Kaufman B, Wildiers H, Friedlander M, Ayoub JP, et al. Veliparib with carboplatin and paclitaxel in BRCA-mutated advanced breast cancer (BROCADE3): a randomised, double-blind, placebo-controlled, phase 3 trial. Lancet Oncol 2020;21(10):1269—82.

[74] Gudmundsdottir K, Ashworth A. The roles of BRCA1 and BRCA2 and associated proteins in the maintenance of genomic stability. Oncogene 2006;25(43):5864—74.

[75] Pellegrino B, Mateo J, Serra V, Balmana J. Controversies in oncology: are genomic tests quantifying homologous recombination repair deficiency (HRD) useful for treatment decision making? ESMO Open 2019;4(2):e000480.

[76] Morgan RD, Clamp AR, Evans DGR, Edmondson RJ, Jayson GC. PARP inhibitors in platinum-sensitive high-grade serous ovarian cancer. Cancer Chemother Pharmacol 2018;81(4):647—58.

[77] Wooster R, Weber BL. Breast and ovarian cancer. N Engl J Med 2003;348 (23):2339—47.

[78] Pilie PG, Gay CM, Byers LA, O'Connor MJ, Yap TA. PARP inhibitors: extending benefit beyond BRCA-mutant cancers. Clin Cancer Res 2019;25(13):3759—71.

[79] Caracciolo D, Riillo C, Arbitrio M, Di Martino MT, Tagliaferri P, Tassone P. Error-prone DNA repair pathways as determinants of immunotherapy activity: an emerging scenario for cancer treatment. Int J Cancer 2020;147(10):2658—68.

[80] Caracciolo D, Riillo C, Di Martino MT, Tagliaferri P, Tassone P. Alternative non-homologous end-joining: error-prone DNA repair as cancer's Achilles' Heel. Cancers (Basel) 2021;13(6).

[81] Hanahan D, Weinberg RA. Hallmarks of cancer: the next generation. Cell 2011;144 (5):646—74.

[82] You JS, Jones PA. Cancer genetics and epigenetics: two sides of the same coin? Cancer Cell 2012;22(1):9—20.

[83] Chiba T, Marusawa H, Ushijima T. Inflammation-associated cancer development in digestive organs: mechanisms and roles for genetic and epigenetic modulation. Gastroenterology 2012;143(3):550—63.

[84] Luchini C, Bibeau F, Ligtenberg MJL, Singh N, Nottegar A, Bosse T, et al. ESMO recommendations on microsatellite instability testing for immunotherapy in cancer, and its relationship with PD-1/PD-L1 expression and tumour mutational burden: a systematic review-based approach. Ann Oncol 2019;30(8):1232—43.

[85] Lee V, Murphy A, Le DT, Diaz Jr. LA. Mismatch repair deficiency and response to immune checkpoint blockade. Oncologist 2016;21(10):1200—11.

[86] Seth S, Ager A, Arends MJ, Frayling IM. Lynch syndrome—cancer pathways, heterogeneity and immune escape. J Pathol 2018;246(2):129—33.

[87] Lynch HT, Snyder CL, Shaw TG, Heinen CD, Hitchins MP. Milestones of lynch syndrome: 1895—2015. Nat Rev Cancer 2015;15(3):181—94.

[88] Lawes DA, SenGupta S, Boulos PB. The clinical importance and prognostic implications of microsatellite instability in sporadic cancer. Eur J Surg Oncol 2003;29(3):201—12.

[89] Ligtenberg MJ, Kuiper RP, Chan TL, Goossens M, Hebeda KM, Voorendt M, et al. Heritable somatic methylation and inactivation of MSH2 in families with Lynch syndrome due to deletion of the 3' exons of TACSTD1. Nat Genet 2009;41 (1):112—17.

[90] Glaire MA, Brown M, Church DN, Tomlinson I. Cancer predisposition syndromes: lessons for truly precision medicine. J Pathol 2017;241(2):226—35.

[91] Jiricny J. Postreplicative mismatch repair. Cold Spring Harb Perspect Biol 2013;5(4): a012633.

[92] Gavin PG, Colangelo LH, Fumagalli D, Tanaka N, Remillard MY, Yothers G, et al. Mutation profiling and microsatellite instability in stage II and III colon cancer: an assessment of their prognostic and oxaliplatin predictive value. Clin Cancer Res 2012;18(23):6531—41.

[93] Roth AD, Delorenzi M, Tejpar S, Yan P, Klingbiel D, Fiocca R, et al. Integrated analysis of molecular and clinical prognostic factors in stage II/III colon cancer. J Natl Cancer Inst 2012;104(21):1635—46.

[94] Lochhead P, Kuchiba A, Imamura Y, Liao X, Yamauchi M, Nishihara R, et al. Microsatellite instability and BRAF mutation testing in colorectal cancer prognostication. J Natl Cancer Inst 2013;105(15):1151—6.

[95] Sinicrope FA, Mahoney MR, Smyrk TC, Thibodeau SN, Warren RS, Bertagnolli MM, et al. Prognostic impact of deficient DNA mismatch repair in patients with stage III colon cancer from a randomized trial of FOLFOX-based adjuvant chemotherapy. J Clin Oncol 2013;31(29):3664—72.

[96] Klingbiel D, Saridaki Z, Roth AD, Bosman FT, Delorenzi M, Tejpar S. Prognosis of stage II and III colon cancer treated with adjuvant 5-fluorouracil or FOLFIRI in relation to microsatellite status: results of the PETACC-3 trial. Ann Oncol 2015;26 (1):126—32.

[97] Andre T, de Gramont A, Vernerey D, Chibaudel B, Bonnetain F, Tijeras-Raballand A, et al. Adjuvant fluorouracil, leucovorin, and oxaliplatin in stage II to III colon cancer: updated 10-year survival and outcomes according to braf mutation and mismatch repair status of the MOSAIC study. J Clin Oncol 2015;33(35):4176—87.

[98] Dienstmann R, Mason MJ, Sinicrope FA, Phipps AI, Tejpar S, Nesbakken A, et al. Prediction of overall survival in stage II and III colon cancer beyond TNM system: a retrospective, pooled biomarker study. Ann Oncol 2017;28(5):1023−31.

[99] Hutchins G, Southward K, Handley K, Magill L, Beaumont C, Stahlschmidt J, et al. Value of mismatch repair, KRAS, and BRAF mutations in predicting recurrence and benefits from chemotherapy in colorectal cancer. J Clin Oncol 2011;29 (10):1261−70.

[100] Samowitz WS, Sweeney C, Herrick J, Albertsen H, Levin TR, Murtaugh MA, et al. Poor survival associated with the BRAF V600E mutation in microsatellite-stable colon cancers. Cancer Res 2005;65(14):6063−9.

[101] Innocenti F, Ou FS, Qu X, Zemla TJ, Niedzwiecki D, Tam R, et al. Mutational analysis of patients with colorectal cancer in CALGB/SWOG 80405 identifies new roles of microsatellite instability and tumor mutational burden for patient outcome. J Clin Oncol 2019;37(14):1217−27.

[102] Venderbosch S, Nagtegaal ID, Maughan TS, Smith CG, Cheadle JP, Fisher D, et al. Mismatch repair status and BRAF mutation status in metastatic colorectal cancer patients: a pooled analysis of the CAIRO, CAIRO2, COIN, and FOCUS studies. Clin Cancer Res 2014;20(20):5322−30.

[103] Des Guetz G, Uzzan B, Nicolas P, Schischmanoff O, Morere JF. Microsatellite instability: a predictive marker in metastatic colorectal cancer? Target Oncol 2009;4 (1):57−62.

[104] Sargent DJ, Marsoni S, Monges G, Thibodeau SN, Labianca R, Hamilton SR, et al. Defective mismatch repair as a predictive marker for lack of efficacy of fluorouracil-based adjuvant therapy in colon cancer. J Clin Oncol 2010;28 (20):3219−26.

[105] Le DT, Kim TW, Van Cutsem E, Geva R, Jager D, Hara H, et al. Phase II open-label study of pembrolizumab in treatment-refractory, microsatellite instability-high/mismatch repair-deficient metastatic colorectal cancer: KEYNOTE-164. J Clin Oncol 2020;38(1):11−19.

[106] Azad NS, Gray RJ, Overman MJ, Schoenfeld JD, Mitchell EP, Zwiebel JA, et al. Nivolumab is effective in mismatch repair-deficient noncolorectal cancers: results from arm Z1D-A subprotocol of the NCI-MATCH (EAY131) study. J Clin Oncol 2020;38(3):214−22.

[107] Prasad V, Kaestner V, Mailankody S. Cancer drugs approved based on biomarkers and not tumor type-FDA approval of pembrolizumab for mismatch repair-deficient solid cancers. JAMA Oncol 2018;4(2):157−8.

[108] Goodman AM, Kato S, Bazhenova L, Patel SP, Frampton GM, Miller V, et al. Tumor mutational burden as an independent predictor of response to immunotherapy in diverse cancers. Mol Cancer Ther 2017;16(11):2598−608.

[109] Rosenberg JE, Hoffman-Censits J, Powles T, van der Heijden MS, Balar AV, Necchi A, et al. Atezolizumab in patients with locally advanced and metastatic urothelial carcinoma who have progressed following treatment with platinum-based chemotherapy: a single-arm, multicentre, phase 2 trial. Lancet. 2016;387 (10031):1909−20.

[110] Rizvi NA, Hellmann MD, Snyder A, Kvistborg P, Makarov V, Havel JJ, et al. Cancer immunology. Mutational landscape determines sensitivity to PD-1 blockade in non-small cell lung cancer. Science 2015;348(6230):124−8.

[111] Snyder A, Makarov V, Merghoub T, Yuan J, Zaretsky JM, Desrichard A, et al. Genetic basis for clinical response to CTLA-4 blockade in melanoma. N Engl J Med 2014;371(23):2189−99.

[112] Cancer Genome Atlas N. Comprehensive molecular characterization of human colon and rectal cancer. Nature 2012;487(7407):330−7.

[113] Eizuka M, Sugai T, Habano W, Uesugi N, Takahashi Y, Kawasaki K, et al. Molecular alterations in colorectal adenomas and intramucosal adenocarcinomas defined by high-density single-nucleotide polymorphism arrays. J Gastroenterol 2017;52(11):1158−68.

[114] Guinney J, Dienstmann R, Wang X, de Reynies A, Schlicker A, Soneson C, et al. The consensus molecular subtypes of colorectal cancer. Nat Med 2015;21 (11):1350−6.

[115] Overman MJ, McDermott R, Leach JL, Lonardi S, Lenz HJ, Morse MA, et al. Nivolumab in patients with metastatic DNA mismatch repair-deficient or microsatellite instability-high colorectal cancer (CheckMate 142): an open-label, multicentre, phase 2 study. Lancet Oncol 2017;18(9):1182−91.

[116] Morse MA, Overman MJ, Hartman L, Khoukaz T, Brutcher E, Lenz HJ, et al. Safety of nivolumab plus low-dose ipilimumab in previously treated microsatellite instability-high/mismatch repair-deficient metastatic colorectal cancer. Oncologist 2019;24(11):1453−61.

[117] Andre T, Shiu KK, Kim TW, Jensen BV, Jensen LH, Punt C, et al. Pembrolizumab in microsatellite-instability-high advanced colorectal cancer. N Engl J Med 2020;383(23):2207−18.

[118] Zhang L, Peng Y, Peng G. Mismatch repair-based stratification for immune checkpoint blockade therapy. Am J Cancer Res 2018;8(10):1977−88.

[119] Goel A, Nagasaka T, Hamelin R, Boland CR. An optimized pentaplex PCR for detecting DNA mismatch repair-deficient colorectal cancers. PLoS One 2010;5(2): e9393.

[120] Vanderwalde A, Spetzler D, Xiao N, Gatalica Z, Marshall J. Microsatellite instability status determined by next-generation sequencing and compared with PD-L1 and tumor mutational burden in 11,348 patients. Cancer Med 2018;7(3):746−56.

[121] Gan C, Love C, Beshay V, Macrae F, Fox S, Waring P, et al. Applicability of next generation sequencing technology in microsatellite instability testing. Genes (Basel) 2015;6(1):46−59.

[122] Timmermann B, Kerick M, Roehr C, Fischer A, Isau M, Boerno ST, et al. Somatic mutation profiles of MSI and MSS colorectal cancer identified by whole exome next generation sequencing and bioinformatics analysis. PLoS One 2010;5 (12):e15661.

[123] Salipante SJ, Scroggins SM, Hampel HL, Turner EH, Pritchard CC. Microsatellite instability detection by next generation sequencing. Clin Chem 2014;60 (9):1192−9.

[124] Hempelmann JA, Scroggins SM, Pritchard CC, Salipante SJ. MSIplus for integrated colorectal cancer molecular testing by next-generation sequencing. J Mol Diagn 2015;17(6):705−14.

[125] Campbell BB, Light N, Fabrizio D, Zatzman M, Fuligni F, de Borja R, et al. Comprehensive analysis of hypermutation in human. Cancer Cell 2017;171 (5):1042−56 e10.

CHAPTER 2

Liquid biopsy: a right tool in a right context?

M. La Mantia[1,*], S. Cutaia[1,*], V. Gristina[1,*], A. Galvano[1], E. Capoluongo[2,3], C. Rolfo[4], U. Malapelle[5], L. Incorvaia[1], G. Badalamenti[1], A. Russo[1,†] and V. Bazan[6,†]

[1]Department of Surgical, Oncological, and Oral Sciences, University of Palermo, Palermo, Italy
[2]Department of Molecular Medicine and Medical Biotechnology, University of Naples Federico II, Naples, Italy
[3]CEINGE, Advances Biotecnologies, Naples, Italy
[4]Center for Thoracic Oncology, Tisch Cancer Institute, Mount Sinai Medical System & Icahn School of Medicine, Mount Sinai, NY, United States
[5]Department of Public Health, University of Naples Federico II, Naples, Italy
[6]Department of Biomedicine, Neuroscience and Advanced Diagnostics (Bi.N.D.), University of Palermo, Palermo, Italy

SUBCHAPTER 2.1
Liquid biopsy in NSCLC

Learning objectives

By the end of the chapter the reader will:
- have a deep knowledge of the application of liquid biopsy,
- know the limitations of liquid biopsy, and
- have learned clinical application of liquid biopsy in different settings of non—small cell lung cancer (NSCLC).

In the last few years, thanks to scientific advances, especially in the field of molecular biology, the management of cancer patients has changed or rather has been revolutionized. It was concluded that to better manage the diagnostic and therapeutic process it is now the comparison between different professional figures, each indispensable with specific skills [1].

* La Mantia M, Cutaia S, and Gristina V. should be considered equally co-first authors.
† Russo A and Bazan V should be considered equally co-last authors.

Liquid Biopsy
DOI: https://doi.org/10.1016/B978-0-12-822703-9.00013-2

© 2023 Elsevier Inc.
All rights reserved.

Too often we have found ourselves with a wealth of information from molecular analyzes that could not be handled. To overcome this problem, the Tumor Board was born which has now expanded to become a Molecular Tumor Board (MTB) comprising the oncologist, the hematologist, the pathologist, the molecular biologist, the clinical biologist, the geneticist, the surgeon, and the bioinformatician [2].

All this allows us to carry the concept forward and the modus operandi of precision medicine according to which by identifying a specific molecular driver one can act against or in favor of it from a therapeutic point of view.

Molecular drivers are usually identified by tumor tissue analysis: one example is the identification of *RAS (KRAS/NRAS)* or *BRAF* mutations in patients with metastatic colon cancer who are not treated with EGFR inhibitor drugs (Fig. 2.1.1) [3,4].

Another example is that related to the identification of molecular drivers in lung cancer such as epidermal growth factor receptor (EGFR)

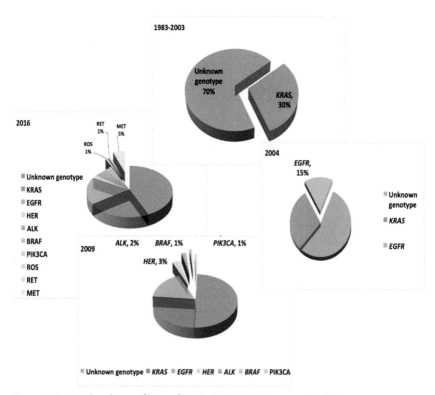

Figure 2.1.1 Molecular profiling of NSCLC. *NSCLC*, Non−small cell lung cancer.

mutations. Most EGFR mutations in NSCLC occur in EGFR exons 18−21 and in patients who have never been smokers or light smokers, but observation is not exclusive. Gefitinib and erlotinib are first-generation reversible EGFR tyrosine kinase inhibitors (EGFR TKIs), afatinib and dacomitinib are second-generation irreversible EGFR TKIs while Osimertinib is a third-generation inhibitor [5].

EGRF analysis is usually performed on tissue but there are other cases where circulating tumor DNA (ctDNA) taken by liquid biopsy can be used as a tissue surrogate.

Among other things, the literature data confirmed that cancer cells change over time, they change, acquiring characteristics that make them resistant to therapies.

Very often the tissue on which to perform all the necessary molecular investigations is not always available and it is from this need that the idea of using liquid biopsy arises.

Liquid biopsy is a minimally or noninvasive technique for finding that circulates tumor components in biofluids. Generally, blood is used but any biological fluid can be used such as urine, pleural effusion, saliva, etc.

Cells, including cancer cells, release DNA fragments called ctDNA, which represents a fraction of the circulating free DNA (cell-free DNA, cfDNA) and which represents the main biomarker user, through necrosis and apostosis mechanisms; it is also possible to use exosomes (EXO), that is, subcellular structures encapsulated in the membrane containing proteins and nucleic acids released by tumor cells, microRNA (miRNA), platelets. All cancers, both primary and metastatic sites, are also capable of exfoliating circulating cancer cells [6,7].

As you can imagine, liquid biopsy has several advantages over tissue biopsy, and above all, it is a noninvasive and repeatable procedure over time. This allows, while maintaining low costs, to have a snapshot of the molecular structure of the disease whenever necessary, with consequences on the therapeutic and/or follow-up side. It, therefore, provides information on the heterogeneity of the disease and its changes over time. This technique is not free from problems related to the management of the biological sample, the analyzes that can be performed, and the absence of a standardized procedure at the time [8].

The development of this technique has made its way thanks to the advent of next-generation sequencing technologies (NGS) that have proved valid in the search for new biomarkers. Surely one of the fields of application in which liquid biopsy is involved is that of NSCLC (stage

IIIB-C and IV) to detect activating and resistance mutations of EGFR in two different clinical scenarios:

- At the time of diagnosis in naive patients when tissue is not available
- At the time of disease progression according to RECIST criteria after treatment with TKI

In the first clinical scenario, a liquid biopsy is a valid option when no tissue samples are available (if it is not possible to perform a new biopsy or if the material is not sufficient) to evaluate the mutational status of EGFR and then-candidate the patient to treatment with EGFR-TKI [9].

In the second clinical setting, ctDNA analysis could be used to detect resistance mutations to first and second-generation EGRF inhibitors.

The liquid biopsy comes into play as a first approach to detecting the p.T790M resistance mutation.

This mutation is a mutation with a dual role: resistance for first and second-generation EGFR TKIs but activating for third-generation EGRF TKIs. If the analysis on the liquid biopsy is positive for the p.T790M mutation, patients can be treated with third-generation TKI, if instead, the test is negative, it is advisable to proceed with the search for the mutation on a tissue biopsy because about 30% of negative liquid biopsy tests for p. T790M are false negative. Indeed, it has been found that the outcome of the liquid biopsy could be influenced by the extent of the disease, by the sites of metastasis (intrathoracic versus extrathoracic disease) that affect the diagnostic accuracy of ctDNA analysis in detecting EGFR mutations [10].

With the new indications regarding the use of Osimentinib upfront according to the data from the FLAURA study, the detection of the p. T790M mutation in this context becomes of secondary importance. However, monitoring of the T790M mutation could be important in predicting the radiological progression of the disease and in explaining further resistance mechanisms [11].

The use of liquid biopsy is also gaining ground in the context of other pathologies thanks to the identification of molecular drivers, through the analysis of ctDNA with NGS. Given the complexity resulting from these analyzes, it is important that the information is analyzed and filtered within the MTB.

Emerging applications of liquid biopsy are found instead in the context of other oncological pathologies such as colorectal carcinoma, breast carcinoma, etc.

In the field of colorectal cancer, several studies are trying to attribute a precise role to the quantization of ctDNA as a prognostic factor in the early stages of the disease to identify Minimal Residual Disease and thus help in estimating the risk of relapse, while in the context of the advanced states, the studies are focusing on the analysis of RAS and BRAF mutations for the choice and monitoring of target therapies [12,13].

On the one hand, this could serve to identify that share of patients with more aggressive disease and therefore be able, from the beginning, to intensify treatment and follow-up programs, while on the other it would offer an alternative to tissue analysis. and would allow for the monitoring of RAS and BRAF mutations over time.

As far as breast cancer is concerned, we know that it is an extremely heterogeneous disease, and its evolutionary history as well as its treatments are linked to a series of factors expressed as the state of hormone receptors for estrogen and progesterone, human EGFR 2, Ki 67, degree of cell differentiation and histotypes. The set of these characteristics with the size of the tumor, the lymph node involvement, and the presence or absence of distant metastases defines its characteristics and stage [14].

Also in this context, liquid biopsy is emerging in the use of biomarkers, such as ctDNA, as predictors of response to treatments or early indicators of disease progression. An important role is that they play in the identification of mutations in the ESR1 gene, have a predictive role of resistance to treatment with CDK4/6 inhibitors, and also in the identification of mutations in the PIK3CA gene. In 2019 the FDA approved treatment with alpelisib, an inhibitor of the PIK3CA gene, for patients with PIK3CA-mutated metastatic breast cancer, and mutations in this gene can be looked for in both blood tissue and liquid biopsy [15−20].

It is clear how liquid biopsy is making its way into all oncological pathologies but another context in which this technique is emerging is that of screening and early diagnosis.

Identification of tumor-specific mutations by ctDNA analysis can act as a new type of cancer biomarker [21−23].

Liquid biopsy is already used in the field of cancer prevention, just think of the identification of pathogenic sequence variants germline in the genes BRCA 1/2 that predispose to the onset of the breast, ovary, prostate, and pancreas. Looking back then, the first uses of liquid biopsy are not that recent [24−26].

Knowledge of this information allows BRCA mutation carriers to take targeted cancer prevention measures that also include surgical options.

Liquid biopsy, therefore, represents a new channel that can be exploited in the management of cancer patients: it is a method capable of revolutionizing clinical practice. In the next few years, it will certainly be possible to understand the most suitable context for this method both in the context of oncological pathologies and in that of early diagnosis and screening.

It will be necessary to learn to understand and filter the information, thanks to the involvement of a team of experts, which can be drawn from this type of exam.

2.1.1 Expert opinion

The unit in precision medicine is a "biomarker ensemble" that includes a predictive biomarker, which plays a crucial role in the carcinogenesis pathway; a diagnostic assay, to identify the biomarker status; and final a therapeutic agent, intended to be more effective for selected patients. With the complexity of the molecular profiling of solid tumors, we understand now that carcinogenesis is based on dynamic molecular networks rather than on a direct connection between genotype and phenotype.

2.1.2 Key points

- Liquid biopsy is a minimally or noninvasive technique for finding that circulates tumor components in biofluids.
- Monitoring the T790M mutation could be important in predicting the radiological progression of the disease and in explaining further resistance mechanisms.
- Liquid biopsy is emerging in the use of biomarkers, such as ctDNA, as predictors of response to treatments or early indicators of disease progression.
- Liquid biopsy, therefore, represents a new channel that can be exploited in the management of cancer patients: it is a method capable of revolutionizing clinical practice.

2.1.3 Summary of clinical recommendations

- AIOM 7 http://www.aiom.it/professionisti/documenti-scientifici/linee-guida/1,413,1,#TopList
- ESMO7 http://www.esmo.org/Guidelines/Lung-and-Chest-Tumours
- NCCN7 https://www.nccn.org/professionals/physician_gls/pdf/nscl.pdf https://www.nccn.org/professionals/physician_gls/pdf/sclc.pdf

Acknowledgments

M. La Mantia, S. Cutaia, and V. Gristina contributed to the current work under the Doctoral Program in Experimental Oncology and Surgery, University of Palermo.

References

[1] Fennell ML, Prabhu Das I, Clauser S, Petrelli N, Salner A. The organization of multidisciplinary care teams: modeling internal and external influences on cancer care quality. JNCI Monogr 2010;2010(40):72−80. Available from: https://doi.org/10.1093/jncimonographs/lgq010.

[2] Schwaederle M, Parker BA, Schwab RB, Fanta PT, Boles SG, Daniels GA, et al. Molecular tumor board: the University of California San Diego Moores cancer center experience. Oncologist 2014;19(6):631−6. Available from: https://doi.org/10.1634/theoncol-ogist.2013-0405.

[3] Pietrantonio F, Petrelli F, Coinu A, et al. Predictive role of BRAF mutations in patients with advanced colorectal cancer receiving cetuximab and panitumumab: a meta-analysis. Eur J Cancer 2015;51:587−94.

[4] Douillard J-Y, Siena S, Cassidy J, et al. Randomized, phase III trial of Panitumumab with Infusional fluorouracil, Leucovorin, and Oxaliplatin (FOLFOX4) vs FOLFOX4 alone as first- line treatment in patients with previously untreated metastatic colorectal cancer: the PRIME study. J Clin Oncol 2010;28:4697−705.

[5] Offin M, Rizvi H, Tenet M, et al. Tumor mutation burden and efficacy of EGFR-tyrosine kinase inhibitors in patients with *EGFR*-mutant lung cancers. Clin Cancer Res 2019;25(3):1063−9. Available from: https://doi.org/10.1158/1078-0432.CCR-18-1102.

[6] Perakis S, Speicher MR. Emerging concepts in liquid biopsies. BMC Med 2017;15(1):75. Available from: https://doi.org/10.1186/s12916-017-0840-6 Published 2017 Apr 6.

[7] Palmirotta R, Lovero D, Cafforio P, et al. Liquid biopsy of cancer: a multimodal diagnostic tool in clinical oncology. Ther Adv Med Oncol 2018;10. Available from: https://doi.org/10.1177/1758835918794630 1758835918794630. Published 2018 Aug 29.

[8] Russo A, Giordano A, Rolfo C, et al. Liquid biopsy in cancer patients: the hand lens to investigate tumor evolution. Current Clinical Pathology. Springer International Publishing; 2017.

[9] Rolfo C, Mack PC, Scagliotti GV, et al. Liquid biopsy for advanced non-small cell lung cancer (NSCLC): a statement paper from the IASLC. J Thorac Oncol 2018;13:1248−68.

[10] Passiglia F, Rizzo S, Rolfo C, et al. Metastatic site location influences the diagnostic accuracy of ctDNA EGFR-mutation testing in NSCLC patients: a pooled analysis. Curr Cancer Drug Targets 2018;18:697−705.

[11] Oxnard GR, Paweletz CP, Kuang Y, et al. Noninvasive detection of response and resistance in EGFR-mutant lung cancer using quantitative next-generation genotyping of cell-free plasma DNA. Clin Cancer Res 2014;20:1698−705.

[12] Reinert T, Henriksen TV, Christensen E, et al. Analysis of plasma cell-free DNA by ultradeep sequencing in patients with stages I to III colorectal cancer. JAMA Oncol 2019;5(8):1124.

[13] Normanno N, Cervantes A, Ciardiello F, et al. The liquid biopsy in the management of colorectal cancer patient: current applications and future scenarios. Cancer Treat Rev 2018;70:1−8.

[14] Gradishar WJ, Anderson BO, Abraham J, Aft R, Agnese D, Allison KH, et al. Breast Cancer, Version 3.2020, NCCN Clinical Practice Guidelines in Oncology. J Natl

Compr Canc Netw 2020;18(4):452–78. Available from: https://doi.org/10.6004/jnccn.2020.0016. PMID: 32259783.

[15] Bidard F.C., Pistilli B., Dalenc F., et al. Circulating ESR1 mutation detection rate and early decrease under first line aromatase inhibitor and palbociclib in the PADA-1 trial (UCBG-GINECO). Cancer Research 2019;79 (Suppl. 4). Available from: https://doi.org/10.1158/1538-7445.SABCS18-PD2-06.

[16] Andre F, Ciruelos E, Rubovszky G, et al. Alpelisib for PIK3CA-mutated, hormone receptor-positive advanced breast cancer. N Engl J Med 2019;380:1929–40.

[17] US Preventive Services Task Force, Owens DK, Davidson KW, Krist AH, Barry MJ, Cabana M, et al. Risk assessment, genetic counseling, and genetic testing for BRCA-related cancer: US preventive services task force recommendation statement. JAMA 2019;322(7):652–65. Available from: https://doi.org/10.1001/jama.2019.10987 Erratum in: JAMA. 2019 Nov 12;322(18):1830. PMID: 31429903.

[18] Incorvaia L, Russo A, Cinieri S. The molecular tumor board: a tool for the governance of precision oncology in the real world. Tumori 2021. Available from: https://doi.org/10.1177/03008916211062266 Epub ahead of print. PMID: 34918610.

[19] Russo A, Incorvaia L, Del Re M, Malapelle U, Capoluongo E, Gristina V, et al. The molecular profiling of solid tumors by liquid biopsy: a position paper of the AIOM-SIAPEC-IAP-SIBioC-SIC-SIF Italian Scientific Societies. ESMO Open 2021;6 (3):100164. Available from: https://doi.org/10.1016/j.esmoop.2021.100164 Epub 2021 Jun 3. PMID: 34091263.

[20] Russo A, Incorvaia L, Malapelle U, Del Re M, Capoluongo E, Vincenzi B, et al. The tumor-agnostic treatment for patients with solid tumors: a position paper on behalf of the AIOM- SIAPEC/IAP-SIBioC-SIF Italian Scientific Societies. Crit Rev Oncol Hematol 2021;165:103436. Available from: https://doi.org/10.1016/j.critrevonc.2021.103436 Epub 2021 Aug 8. PMID: 34371157.

[21] Russo A, Incorvaia L, Capoluongo E, Tagliaferri P, Galvano A, Del Re M, et al. The challenge of the Molecular Tumor Board empowerment in clinical oncology practice: a position paper on behalf of the AIOM- SIAPEC/IAP-SIBioC-SIC-SIF-SIGU-SIRM Italian Scientific Societies. Crit Rev Oncol Hematol 2022;169:103567. Available from: https://doi.org/10.1016/j.critrevonc.2021.103567 Epub 2021 Dec 8. PMID: 34896250.

[22] Passiglia F, Galvano A, Gristina V, et al. Is there any place for PD-1/CTLA-4 inhibitors combination in the first-line treatment of advanced NSCLC?-a trial-level meta-analysis in PD-L1 selected subgroups. Transl Lung Cancer Res 2021;10(7):3106–19. Available from: https://doi.org/10.21037/tlcr-21-52.

[23] Gristina V, La Mantia M, Iacono F, Galvano A, Russo A, Bazan V. The emerging therapeutic landscape of alk inhibitors in non-small cell lung cancer. Pharm (Basel) 2020;13 (12):474. Available from: https://doi.org/10.3390/ph13120474 Published 2020 Dec 18.

[24] Pisapia P, Pepe F, Gristina V, et al. A narrative review on the implementation of liquid biopsy as a diagnostic tool in thoracic tumors during the COVID-19 pandemic. Mediastinum 2021;5:27. Available from: https://doi.org/10.21037/med-21-9 Published 2021 Sep 25.

[25] Incorvaia L, Fanale D, Bono M, et al. BRCA1/2 pathogenic variants in triple-negative vs luminal-like breast cancers: genotype-phenotype correlation in a cohort of 531 patients. Ther Adv Med Oncol 2020;12. Available from: https://doi.org/10.1177/1758835920975326 1758835920975326. Published 2020 Dec 16.

[26] Passiglia F, Rizzo S, Rolfo C, Galvano A, Bronte E, Incorvaia L, et al. Metastatic site location influences the diagnostic accuracy of ctDNA EGFR- mutation testing in NSCLC patients: a pooled analysis. Curr Cancer Drug Targets 2018;18(7):697–705. Available from: https://doi.org/10.2174/1568009618666180308125110. PMID: 29521235.

Further reading

Incorvaia L, Russo A, Cinieri S. The molecular tumor board: a tool for the governance of precision oncology in the real world. Tumori 2021. Available from: https://doi.org/10.1177/03008916211062266 Epub ahead of print. PMID.

Russo A, Incorvaia L, Del Re M, Malapelle U, Capoluongo E, Gristina V, et al. The molecular profiling of solid tumors by liquid biopsy: a position paper of the AIOM-SIAPEC-IAP-SIBioC-SIC-SIF Italian Scientific Societies. ESMO Open 2021;6(3):100164. Available from: https://doi.org/10.1016/j.esmoop.2021.100164 Epub 2021 Jun 3. PMID.

Russo A, Incorvaia L, Malapelle U, Del Re M, Capoluongo E, Vincenzi B, et al. The tumor-agnostic treatment for patients with solid tumors: a position paper on behalf of the AIOM- SIAPEC/IAP-SIBioC-SIF Italian Scientific Societies. Crit Rev Oncol Hematol 2021;165:103436. Available from: https://doi.org/10.1016/j.critre-vonc.2021.103436 Epub 2021 Aug 8. PMID.

SUBCHAPTER 2.2

The role of mutated ctDNA in nonmalignant lesions: challenging aspects in liquid biopsy implementation

V. Gristina[1], U. Malapelle[2], P. Pisapia[2], M. La Mantia[1], R. Scalia[1], M. Greco[1],

L. Incorvaia[1], G. Badalamenti[1], A. Galvano[1], A. Russo[1] and V. Bazan[3,†]

[1]Department of Surgical, Oncological, and Oral Sciences, University of Palermo, Palermo, Italy
[2]Department of Public Health, University of Naples Federico II, Naples, Italy
[3]Department of Biomedicine, Neuroscience and Advanced Diagnostics (Bi.N.D.), University of Palermo, Palermo, Italy

Learning objectives

By the end of the chapter the reader will:
- have learned the basic concepts of clinical employment of liquid biopsy;
- have deep knowledges on circulating tumor cells, circulating nucleic acids, and micro-extracellular vesicles; and
- have learned the clinical application of liquid biopsy in non—small cell lung cancer (NSCLC).

In advanced-stage cancer patients, the analysis of circulating tumor DNA (ctDNA) represents a useful tool to perform molecular analysis for

[†] Russo A and Bazan V should be considered equally co-last authors.

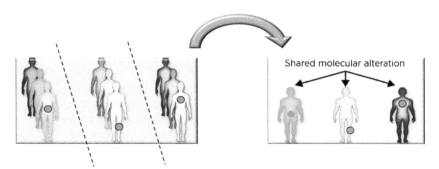

Figure 2.2.1 Integration of clinical-translational strategies into precision medicine drug development.

diagnostic, predictive and prognostic purposes, even in absence of tissue specimens [27,28]. Unfortunately, discordant results between tissue and ctDNA analysis have emerged. In particular, ctDNA lower sensitivity compared to tissue genotyping may be due to a variable shed of tumor DNA into the plasma, resulting in false-negative ctDNA results [29]. Another setting in which discrepancies may arise is represented by the drug resistance due to tumor heterogeneity [30,31] (Fig. 2.2.1). Beyond these established biological factors, technical factors may be an underappreciated source of erroneous plasma sequencing [32−34].

Besides this evidence, other mechanisms can be involved in false-positive results on ctDNA extracted from plasma analysis [35,36]. Both germline and noncancerous somatic variants are also known to confound the analysis and interpretation of plasma NGS results, eventually leading to misleading and harmful test results [37,38]. In this setting, an important mechanism that can lead to false detection of ctDNA mutations is represented by clonal hematopoiesis of indeterminate potential (CIMP) [39,40]. As suggested by the name, this phenomenon is characterized by the presence of mutations in the DNA of hematopoietic stem cells or other early blood cell progenitors [41,42]. This occurrence may determine the detection of noncancer-related mutations in ctDNA [43,44]. As reported in the large experience by Genovese et al., clonal hematopoiesis determined the development of somatic mutations in 10% of the study participants, with increasing frequency with age [45,46]. Due to the high concentration in the bloodstream, peripheral blood cells give an important contribution to the composition of circulating cell-free DNA (ccfDNA). Noteworthy, ctDNA represents a very low part of the total ccfDNA (<0.5%) [47,48]. Clonal hematopoiesis was reported in a not negligible percentage (25%) of patients who harbored a solid tumor [49]. Additionally, Razavi et al. observed that among cancer patients half of the cfDNA genomic alterations were originated in the hematological compartment, proving not to be tumor-derived [50,51]. To overcome this limitation, a high-intensity deep sequencing assay with a targeted approach analysis was performed on matched tumoral tissue, cfDNA, and white blood cells, confirming the presence of CIMP-related alterations in cfDNA from both healthy individuals and cancer patients [50,52]. Of note, not all mutations are interested in this phenomenon. EGFR, V-Raf Murine Sarcoma Viral Oncogene Homolog B (BRAF), and MET Proto-Oncogene, Receptor Tyrosine Kinase (MET) clinical relevant mutations seems to be not associated with clonal hematopoiesis. For this reason, when these alterations are identified in ctDNA extracted from plasma they can be adopted for targeted therapies. Conversely, Kirsten Rat Sarcoma Viral Oncogene Homolog (KRAS) or Tumor Protein P53 (TP53) gene mutations may be associated with clonal hematopoiesis [39,53]. In the study by Hu et al., most Janus Kinase 2 mutations, TP53 mutations, and rare KRAS mutations are derived from clonal hematopoiesis [39,54]. On the other hand, the association of CIMP with cancer risk remains to be determined since an increased risk with myeloid neoplasms as well as cardiovascular diseases has been observed [55–57]. In this vein, further research studies are needed to assess the clinical utility of CIMP as a valid and reliable tool in the decision-making process [58,59].

Finally, other nontumoral somatic genomic alterations arising in other aging tissues that are not limited to the hematopoietic compartment need to be considered when using deep-sequencing NGS approaches, as little is known about endometriosis and aneurysm [38,60]. In this context, future strategies based on tools such as methylation pattern profiling that can identify the source tissue of cfDNA are under investigation and eagerly awaited to be validated for the accurate identification of ctDNA in liquid biopsy [61,62].

Another relevant setting in which ctDNA mutational assessment may play an important role is represented by cancer screening. In a recent experience, Pepe et al. adopted a novel high sensitive real-time polymerase chain reaction (RT-PCR) based device (ColoScape kit, DiaCarta, Richmond, CA) able to detect mutations in Adenomatous Polyposis Coli (APC; codons 1309, 1367, 1450,) KRAS (codons 12 and 13), BRAF (codon 600) and Catenin Beta 1 (CTNNB1; codons 41 and 45) genes in both plasma and formalin-fixed and paraffin-embedded (FFPE) samples [63]. In this pilot study, people with positive detection of blood in the stool (Fecal Immunochemical Test positive [FIT +]) underwent ctDNA analysis, and colonoscopy was adopted as a confirmatory approach (gold standard). To date, this novel approach on ctDNA showed a sensitivity, specificity, positive predictive value, and negative predictive value of 53.8%, 92.3%, 70.0%, and 85.7%, respectively, when considering advanced adenomas [63,64]. As far as colorectal cancer screening is concerned, ctDNA showed a key role as a noninvasive screening tool for asymptomatic people through the evaluation of the methylation status of the Septin 9 (*SEPT9*) gene [65−69].

2.2.1 Expert opinion

One of the crucial and modern clinical employments of liquid biopsy is the tracking of ctDNA by longitudinal blood extractions, that allows for a global, dynamic analysis of tumor genomic landscape. This might endorse in the early detection of resistance mutations emerging during oncological treatment before evidence of clinical or radiological disease progression. These findings could lead the health-care providers in order to avoid keeping the administration of noneffective oncological treatments and develop new therapeutic approaches to empower the clinical outcome of patients.

2.2.2 Key points

- The term liquid biopsy relates to the detection, quantification, and analysis of tumor components that are detected in body biological fluids (plasma, serum, saliva, urine, and effusion liquids).
- Tumor components extracted from body fluids are circulating nucleic acids (circulating tumor DNA, circulating miRNA, and circulating RNA), circulating tumor cells, and extracellular vesicles (EXO and microvesicles).
- The advantage of liquid biopsy is a minimally invasive procedure that captures tumor heterogeneity and real time variations in tumor dynamics.
- Recent advances in biotechnology (i.e., NGS, digital PCR) have empowered the development of liquid biopsy. With increasing modern technologies, liquid biopsy could currently be used in an easy and reproducible way in daily clinical practice.

Acknowledgments

M. La Mantia, and V. Gristina contributed to the current work under the Doctoral Programme in Experimental Oncology and Surgery, University of Palermo.

References

[27] Nacchio M, et al. KRAS mutations testing in non-small cell lung cancer: the role of liquid biopsy in the basal setting. J Thorac Dis 2020;12:3836—43.

[28] Siravegna G, et al. How liquid biopsies can change clinical practice in oncology. Ann Oncol J Eur Soc Med Oncol 2019;30:1580—90.

[29] Sacher AG, et al. Prospective validation of rapid plasma genotyping for the detection of EGFR and KRAS mutations in advanced lung cancer. JAMA Oncol 2016;2:1014—22.

[30] Galvano A, et al. Moving the target on the optimal adjuvant strategy for resected pancreatic cancers: a systematic review with meta-analysis. Cancers (Basel) 2020;12.

[31] Oxnard GR, et al. Association between plasma genotyping and outcomes of treatment with Osimertinib (AZD9291) in advanced non-small-cell lung cancer. J Clin Oncol J Am Soc Clin Oncol 2016;34:3375—82.

[32] Pepe F, et al. Tumor mutational burden on cytological samples: a pilot study. Cancer Cytopathol 2020. Available from: https://doi.org/10.1002/cncy.22400.

[33] Weber S, et al. Technical evaluation of commercial mutation analysis platforms and reference materials for liquid biopsy profiling. Cancers (Basel) 2020;12.

[34] Russo A, et al. The molecular profiling of solid tumors by liquid biopsy: a position paper of the AIOM-SIAPEC-IAP-SIBioC-SIC-SIF Italian Scientific Societies. ESMO open 2021;6:100164.

[35] Mantia ML, Koyyala VPB. The war against coronavirus disease 19 through the eyes of cancer physician: an Italian and Indian young medical oncologist's perspective. Indian J Med Paediatr Oncol 2020;41:305—7.

[36] Incorvaia L, et al. BRCA1/2 pathogenic variants in triple-negative vs luminal-like breast cancer: genotype-phenotype correlation in a cohort of 531 patients. Ther Adv Med Oncol 2020;12 1758835920975326.

[37] Gristina V, et al. Non-small cell lung cancer harboring concurrent EGFR genomic alterations: a systematic review and critical appraisal of the double dilemma. J Mol Pathol 2021;2:173—96.

[38] Bellosillo B, Montagut C. High-accuracy liquid biopsies. Nat Med 2019;25:1820—1.

[39] Hu Y, et al. False-positive plasma genotyping due to clonal hematopoiesis. Clin Cancer Res 2018;24:4437—43.

[40] Scilla KA, Rolfo C. The role of circulating tumor DNA in lung cancer: mutational analysis, diagnosis, and surveillance now and into the future. Curr Treat Options Oncol 2019;20:61.

[41] Steensma DP, et al. Clonal hematopoiesis of indeterminate potential and its distinction from myelodysplastic syndromes. Blood 2015;126:9—16.

[42] Pisapia P, et al. A narrative review on the implementation of liquid biopsy as a diagnostic tool in thoracic tumors during the COVID-19 pandemic. Mediastinum 2021;5(September 2021) *Mediastinum.*

[43] Russo A, et al. The tumor-agnostic treatment for patients with solid tumors: a position paper on behalf of the AIOM- SIAPEC/IAP-SIBioC-SIF Italian Scientific Societies. Crit Rev Oncol Hematol 2021;165:103436.

[44] Russo A, et al. The challenge of the Molecular Tumor Board empowerment in clinical oncology practice: a position paper on behalf of the AIOM- SIAPEC/IAP-SIBioC-SIC-SIF-SIGU-SIRM Italian Scientific Societies. Crit Rev Oncol Hematol 2022;169:103567.

[45] Genovese G, et al. Clonal hematopoiesis and blood-cancer risk inferred from blood DNA sequence. N Engl J Med 2014;371:2477—87.

[46] Incorvaia L, et al. Baseline plasma levels of soluble PD-1, PD-L1, and BTN3A1 predict response to nivolumab treatment in patients with metastatic renal cell carcinoma:

a step toward a biomarker for therapeutic decisions. Oncoimmunology 2020;9 1832348.

[47] Malapelle U, et al. Next generation sequencing techniques in liquid biopsy: focus on non-small cell lung cancer patients. Transl Lung Cancer Res 2016;5:505−10.

[48] Galvano A, et al. Detection of RAS mutations in circulating tumor DNA: a new weapon in an old war against colorectal cancer. A systematic review of literature and meta-analysis. Ther Adv Med Oncol 2019;11 1758835919874653.

[49] Coombs CC, et al. Therapy-related clonal hematopoiesis in patients with non-hematologic cancers is common and associated with adverse clinical outcomes. Cell Stem Cell 2017;21:374−82 e4.

[50] Razavi P, et al. High-intensity sequencing reveals the sources of plasma circulating cell-free DNA variants. Nat Med 2019;25:1928−37.

[51] Beretta G, Capoluongo E, Danesi R, Del Re M, Fassan M, Giuffrè G, et al. Raccomandazioni 2020 per l'esecuzione di Test Molecolari su Biopsia Liquida in Oncologia AIOM. (2020). Available from: https://www.aiom.it/raccomandazioni-2020-per-lesecuzione-di-test-molecolari-su-biopsia-liquida-in-oncologia/.

[52] Passiglia F, et al. Metastatic site location influences the diagnostic accuracy of ctDNA EGFR- mutation testing in NSCLC patients: a pooled analysis. Curr Cancer Drug Targets 2018;18:697−705.

[53] Pisapia P, et al. Next generation diagnostic algorithm in non-small cell lung cancer predictive molecular pathology: the KWAY Italian multicenter cost evaluation study. Crit Rev Oncol Hematol 2021;169:103525.

[54] Galvano A, et al. Analysis of systemic inflammatory biomarkers in neuroendocrine carcinomas of the lung: prognostic and predictive significance of NLR, LDH, ALI, and LIPI score. Ther Adv Med Oncol 2020;12 1758835920942378.

[55] Takahashi K, et al. Preleukaemic clonal haemopoiesis and risk of therapy-related myeloid neoplasms: a case-control study. Lancet Oncol 2017;18:100−11.

[56] Gillis NK, et al. Clonal haemopoiesis and therapy-related myeloid malignancies in elderly patients: a proof-of-concept, case-control study. Lancet Oncol 2017;18:112−21.

[57] Jaiswal S, et al. Clonal hematopoiesis and risk of atherosclerotic cardiovascular disease. N Engl J Med 2017;377:111−21.

[58] Incorvaia L, et al. Challenges and advances for the treatment of renal cancer patients with brain metastases: from immunological background to upcoming clinical evidence on immune-checkpoint inhibitors. Crit Rev Oncol Hematol 2021;163:103390.

[59] Passiglia F, et al. Looking for the best immune-checkpoint inhibitor in pre-treated NSCLC patients: an indirect comparison between nivolumab, pembrolizumab and atezolizumab. Int J Cancer 2018. Available from: https://doi.org/10.1002/ijc.31136.

[60] Galvano A, et al. The prognostic impact of tumor mutational burden (TMB) in the first-line management of advanced non-oncogene addicted non-small-cell lung cancer (NSCLC): a systematic review and meta-analysis of randomized controlled trials. ESMO Open 2021;6.

[61] Moss J, et al. Comprehensive human cell-type methylation atlas reveals origins of circulating cell-free DNA in health and disease. Nat Commun 2018;9:5068.

[62] Novo G, et al. Arterial stiffness: effects of anticancer drugs used for breast cancer women. Front Physiol 2021;12:661464.

[63] Scimia M, et al. Evaluation of a novel liquid biopsy-based ColoScape assay for mutational analysis of colorectal neoplasia and triage of FIT + patients: a pilot study. J Clin Pathol 2018;71:1123−6.

[64] Massihnia D, et al. Triple negative breast cancer: shedding light onto the role of pi3k/akt/mtor pathway. Oncotarget 2016;7:60712−22.

[65] Church TR, et al. Prospective evaluation of methylated SEPT9 in plasma for detection of asymptomatic colorectal cancer. Gut 2014;63:317−25.

[66] Gristina V, et al. The emerging therapeutic landscape of ALK inhibitors in non-small cell lung cancer. Pharm (Basel) 2020;13.

[67] Passiglia F, Galvano A, Gristina V, et al. Is there any place for PD-1/CTLA-4 inhibitors combination in the first-line treatment of advanced NSCLC?-a trial-level meta-analysis in PD-L1 selected subgroups. Transl Lung Cancer Res 2021;10(7):3106−19. Available from: https://doi.org/10.21037/tlcr-21-52.

[68] Incorvaia L, Russo A, Cinieri S. The molecular tumor board: a tool for the governance of precision oncology in the real world. Tumori 2021. Available from: https://doi.org/10.1177/03008916211062266   Epub ahead of print. PMID: 34918610.

[69] Russo A, Incorvaia L, Capoluongo E, Tagliaferri P, Galvano A, Del Re M, et al. The challenge of the Molecular Tumor Board empowerment in clinical oncology practice: a position paper on behalf of the AIOM- SIAPEC/IAP-SIBioC-SIC-SIF-SIGU-SIRM Italian Scientific Societies. Crit Rev Oncol Hematol 2022;169:103567. Available from: https://doi.org/10.1016/j.critrevonc.2021.103567   Epub 2021 Dec 8. PMID: 34896250.

Further reading

Chaudhuri AA, Chabon JJ, Lovejoy AF, et al. Early detection of molecular residual disease in localized lung cancer by circulating tumor DNA profiling. Cancer Discov 2017;12:1394−403.

Cohen JD, Li L, Wang Y, et al. Detection and localization of surgically resectable cancers with a multianalyte blood test. Science. 2018.

Diehl F, Schmidt K, Choti MA, et al. Circulating mutant DNA to assess tumor dynamics. Nat Med 2008;14:985−90.

Siravegna G, Mussolin B, Buscarino M, et al. Clonal evolution and resistance to EGFR blockade in the blood of colorectal cancer patients. Nat Med 2015;21:795−801.

CHAPTER 3

Liquid biopsy: new challenges in the era of immunotherapy and precision oncology NGS and the other faces of molecular biology

F. Buttitta, A. Navicella, R. Rossetti, L. Felicioni and A. Marchetti

Center of Advanced Studies and Technology - University of Chieti, Chieti, Italy

Learning objectives

By the end of the chapter the reader will:

- Have learned the basic concepts of potential clinical application of liquid biopsy.
- Have learned the basic application of next-generation sequencing.

3.1 Tissue or liquid biopsy?

In recent years, important steps forward have been made in various fields of biological research in synergy with progressive technological improvement. In the oncological field, the application of these advancements in clinical practice, together with the design of innovative molecularly targeted drugs, have paved the way for precision medicine. This new pharmacological approach lets to specific oncological treatments for each patient, based on the genetic characteristics of the tumor. The presence of a vast spectrum of gene alterations in neoplastic diseases, that makes the patient eligible for new and effective anticancer treatments, has favored the development of new diagnostic strategies.

Neoplastic tissue is still considered today, among all biological materials, the gold standard for carrying out molecular tests according to the treatment. Genetic analysis provides information both on driver alterations that favored the development of neoplastic growth and on alterations triggering resistance to treatment. Cancer tissue can be obtained through a small biopsy or a larger surgical resection. It should be emphasized that currently targeted molecular therapies are aimed at metastatic or locally

Liquid Biopsy
DOI: https://doi.org/10.1016/B978-0-12-822703-9.00015-6

© 2023 Elsevier Inc.
All rights reserved.

47

advanced malignancies, and therefore not eligible for surgery. In these cases, it is possible at the time of diagnosis to obtain only small samples for tumor characterization.

Such samples are sometimes very small or contain a limited neoplastic component which is not enough for a complete tumor typing or a followed molecular characterization. Several clinical situations do not allow obtaining any histological material: the tumor site may not be accessible to the surgical approach or the patient's clinical status may not allow for surgery. Sometimes, specimens rich in a fibrotic component cannot be diagnostic.

For these reasons, an alternative source of nucleic acids to be analyzed with molecular procedures is necessary, such as circulating tumor cells (CTCs) and circulating cell-free DNA (cfDNA) in advanced and/or metastatic neoplastic patients. In these cases, a blood sample can be considered a substitute for a tissue biopsy and is therefore referred to as a liquid biopsy [1].

Locally advanced or metastatic cancer patients have a variable amount of circulating cancer cells which can now be analyzed with dedicated platforms such as CellSearch. The CellSearch system is the first and the only clinically validated blood test by the Food and Drug Administration (FDA) to detect and enumerate CTCs of epithelial origin (CD45−, EpCAM +, and cytokeratins 8, 18 + and/or 19 +). This approach has clinical value in patients undergoing treatment for metastatic breast, colorectal, or prostate cancer [2,3].

The presence of CTCs characterizes the metastatic process as first highlighted by an Australian pathologist, Thomas Ashworth in 1869. Thomas Ashworth first documented the presence of epithelial cells in the blood compartment of a patient dying of cancer. He raised the possibility that those cells were tumor-derived and potentially explained the presence of multiple tumor metastases [4].

To date, molecular testing on circulating cfDNA has spread and imposed more than the analysis carried out on CTCs. A variable amount of circulating cfDNA, sometimes abundant, accumulates in the blood of neoplastic patients and includes a small fraction of cell-free circulating tumor DNA (cftDNA) [5].

Generally speaking, the liquid biopsy also implies other biological fluids such as urine, ascitic and pleural effusions, saliva, and cerebrospinal fluid that can be used for mutational analysis. Although these biological fluids have limited applications as they provide information for the specific district (e.g., oral cavity, urinary tract, brain, etc.) from which they are taken, they can be useful for answering specific clinical questions. On the contrary, the blood is more representative as it conveys the mutational information from primary and secondary neoplastic lesions.

The presence of circulating cfDNA was reported in human plasma by Mandel and Metais in 1948. Since then, the analysis of circulating cfDNA was totally ignored until the 1970s–1980s [6].

In June 2016, liquid biopsy has received the FDA approval in clinical practice for epidermal growth factor receptor (EGFR) molecular characterization in lung cancer patients, using the Cobas EGFR Mutation Test, the first monomarker test specifically designed for this purpose [7,8]. EGFR mutations, such as deletions in exon 19 and substitution in exon 21, detected on cftDNA, resulted in predicting response to tyrosine kinase inhibitors (TKIs), in patients with non-small cell lung cancer (NSCLC) [9–11].

This approval allows for recovery, after appropriate treatment, for EGFR mutation-positive patients who lack neoplastic tissue or with an exhausted neoplastic specimen.

Several studies have reported excellent concordance between EGFR testing on neoplastic tissue and on circulating cfDNA, especially in lung cancer and colorectal cancer (CRC) patients at an advanced stage (Stage III–IV). On the contrary, starting from circulating cfDNA obtained from early-stage patients a low sensitivity of molecular test has been reported [12–14].

On the DNA extracted from CTCs or from circulating cfDNA, different types of genomic analysis to identify several types of molecular alterations (point mutations, deletions/insertions) can be performed and consequently direct toward the best therapeutic choice for the individual patient (targeted or personalized therapy). Contrary to what occurs for CTCs which, for some types of neoplastic forms, are often present in low numbers, cftDNA in advanced-stage patients is generally appropriate for molecular analysis. It also provides a dynamic picture of the tumor evolution through the identification of further acquired genetic alterations accountable for relapse, progression, and development of drug resistance.

Extensive analysis of many genes by monomarker tests is time-consuming, costly, and requires amounts of cftDNA that cannot easily be obtained with a single draw. A useful strategy to optimize molecular analysis and rapidly obtain information on many altered genes or pathways as a function of targeted treatments is massive parallel sequencing (MPS), also known as next-generation sequencing (NGS). This technological approach also allows, with high throughput platforms, to evaluate the global load of mutations present in tumor tissue, commonly referred to as Tumor Mutational Burden (TMB) or Tumor Mutational Load. TMB may be useful for selecting patients eligible for new immunotherapy treatments [15].

3.2 Next-generation sequencing for identification of gene alterations in liquid biopsy

MPS is a multiparametric sequencing strategy as it allows the simultaneous sequencing of multiple gene traits, even thousands, and the simultaneous sequencing of a few samples as well as many dozens of specimens obtained from matched patients. All the DNA fragments that will be sequenced are part of the so-called DNA library [16].

This sophisticated technology allows the sequencing of an entire exome (whole-genome Sequencing), a sequencing limited to the exonic regions (whole-exome sequencing), or even the sequencing of specific genetic traits for the identification of known and frequent mutations (targeted sequencing). The latter type of NGS is the one that is most widespread [17].

At present, several innovative pharmacological treatments have taken place in clinical practice and require a growing number of molecular tests. Nowadays, these monomarker genetic tests show some disadvantages, and they will soon be used only in particular situations for the following reasons. They do not allow the current necessary high processivity, in terms of many genes analyzed at the same time, and above all, require a high quantity of nucleic acids. On the other hand, NGS technology makes available the simultaneous assessment of a large number of genetic alterations and requires low DNA/RNA input for the preparation of a single NGS DNA library. Different NGS commercial products vary in the number of genes, from tens to hundreds. The selected genes are referred to as the "panel." Tests aimed at first-level diagnostics including a small number of genes, selected from those most frequently altered in specific malignancies and tests based on larger panels or aimed at specific clinical situations are commercially available. Furthermore, the NGS panels can be targeted mainly or exclusively at known mutations such as missense, insertions, deletions, frameshift, and nonsense or alterations such as translocations/fusions and alterations in the number of gene copies (copy number variation, CNV). Depending on the mutational alteration to be detected, the NGS test requires DNA and/or RNA as starting nucleic acid. Newly conceived NGS panels allow for assessing gene translocations/fusions even starting from DNA.

Overall, the different sequencing technologies share three steps: [1] nucleic acids extraction; [2] library preparation and clonal amplification; and [3] sequencing. The sequencing platforms have considerable differences in the output (Gb per run), based on sample volume and coverage; they also differ in sequencing error rates [18,19]. Table 3.1 shows details of the two main sequencing platforms largely used in clinical practice.

For complex clinical situations that require molecular analysis on liquid biopsy is necessary an increased sensitivity and therefore a greater number of DNA fragments sequenced per region (deep sequencing) with respect to that required for NGS analysis on tissue. A variable quantity of circulating cfDNA can be obtained from liquid biopsy, but only a limited quantity is cftDNA. Panels for the analysis of cftDNA, with higher sensitivity, based on a more recent technological strategy referred to as hybridization and capture sequencing, have been developed and commercially available today.

Briefly, targeted sequencing is based on two NGS different chemical strategies underlying the preparation of the genomic library: the amplicon-based and the capture hybridization method.

The first strategy utilizes tens of thousands of primer pairs for the amplification of specific regions of interest. Two PCR steps follow, each characterized by a qualitative evaluation. At the end of the second PCR, the generated amplicons, are grouped in equimolar quantities and get ready for sequencing after binding with the adapters. The amplicon-based sequencing method has some limitations in that it does not allow for the analysis of gene translocations on DNA. Furthermore, as a consequence of

Table 3.1 Illumina and Thermo Fisher sequencing platforms: read length, output range, run time, and overall error rate in comparison.

		Read length	Output range	Run time	Error rate (%)
Illumina	MiSeq	2 × 300 bp	540 Mb–15 Gb	5–65 h	0.2
	NextSeq 500	2 × 150 bp	30–120 Gb	12–30 h	0.2
	HiSeq X	2 × 150 bp	1.6–1.8 Tb	< 3 days	0.1
	NovaSeq	2 × 150 bp	∼ 6 Tb	∼40 h	0.1
Thermo Fisher	IonPGM	up to 400 bp	up to 2 Gb	2–7 h	1
	IonProton	∼200 bp	up to 10 Gb	2–4 h	1

PCR amplification, sequencing errors are piling up and they affect the final result.

The capture sequencing strategy is based on the hybridization of biotinylated DNA or RNA probes with specific regions of interest. To this aim, DNA is fragmented by physical cutting or enzymatic methods and made suitable for sequencing with the addition of specific adapters for the platform used. The capture is obtained within the hybridization solution using probes complementary to the regions of interest. Gene panels based on the capture hybridization method are often used for clinical applications, due to their reproducibility, specificity, and sensitivity.

This strategy allows for excellent coverage of entire genes and therefore can detect known and unknown mutations unlike what occurs for NGS systems based on the amplification of hot spot gene traits. Regarding fusions, it is also possible to identify new variants of any gene present in the panel, but at least one of the fusion partners must be present.

Regarding the sensitivity of the two strategies, the sequencing with genomic panels capture-based allows switching from a sensitivity of approximately 1% of the amplicon-based panels to a sensitivity of 0.1%. and ensures optimal coverage through sequencing of overlapping genomic traits repeated thousands of times [20].

To date, among the NGS sequencing procedures with the highest sensitivity and specificity, we should mention the CAPP-Seq, based on hybridization and capture. This approach joins the enrichment of alleles with in-depth sequencing, for maximum recovery of low-input DNA, and additional adaptive computer algorithms. Biotinylated oligonucleotides called "selectors" are used, recognizing known mutated sequences annotated on public mutational databases [21]. Population-level bioinformatics analysis creates the CAPP-Seq selectors library, which can be used to detect custom biomarkers. The specificity of this technology has been implemented using a method Integrated Digital Error Suppression (iDES) based on barcodes to tag double-stranded DNA alleles along with a suitable bioinformatics pipeline [22]. It has been shown that this procedure is able to detect the mutant allele frequency with a sensitivity close to 100% in patients with NSCLC. This technology has been adopted in Avenio panels (Roche Diagnostics) commercially available [23].

Another method was devised to develop a very sensitive and highly specific sequencing that can be used for the detection of genetic mutations in blood from patients with early-stage cancers. This procedure indicated by the acronym TEC-Seq combines the targeted capture of multiple regions with a bioinformatics pipeline developed to avoid contamination errors and sequencing artifacts [24].

Different commercial panels for circulating cfDNA analysis are currently available for research and diagnostic purposes (Table 3.2).

3.3 Liquid biopsy in monitoring response to therapy

The liquid biopsy represents a useful diagnostic tool not only for the therapeutic choice but also for evaluating response to therapy and the development of resistance mutations. It has been adopted in clinical practice for EGFR-positive lung cancer patients, but nowadays the experience gained in the field of massive sequencing allows us to use it for patients with other neoplastic forms as followed described.

In NSCLC, alterations in the EGFR are the most frequent druggable mutations. These are in-frame deletions of exon 19 (ex19del) and L858R substitution in exon 21. Rare alterations are missense mutations of exon 18, in-frame insertions in exon 19, and point mutations of exons 20 and 21 [25].

TKIs of first-generation (gefitinib, erlotinib) or third-generation (osimertinib) show impressive efficacy in EGFR-positive lung cancer patients. Sometimes a short response is evident. Our and many other studies have shown that, in many patients, the mutational analysis on liquid biopsy revealed an early progressive quantitative decrease in EGFR-mutated alleles [26−29]. This decrease was directly correlated with the percentage tumor shrinkage at 2 months, assessed by Response Evaluation Criteria in Solid Tumors. Based on the greater or lesser decrease, the patients could be categorized as "rapid responders" (decreasing greater than 50% at 14 days) and "slow responders" (decreasing lower than 50% at 14 days). In some patients showing a slow response, the EGFR driver mutation in the plasma was not completely cleared by TKIs treatment of first-second generation and an early increase in the circulating levels of the T790M resistance mutation was observed. Massive sequencing of circulating cfDNA from "slow responder" patients can highlight the specific resistance mechanism as we recognized in a series of EGFR-positive patients, at progression under treatment with osimertinib [30]. Other druggable alterations can be documented as resistance mutations in patients treated with anti-EGFR-TKIs such as those in the BRAF gene. The BRAF gene is mutated in 4% of lung cancer patients. In half of the cases, the genetic

Table 3.2 Multigene panels applied to liquid biopsy in clinical practice: the number of genes analyzed, platforms, type of mutations identified and sample required.

Test	N° genes analyzed	Platform	Type of mutations	Sample required
Avenio ctDNA Targeted Kit (Roche)	17/77/197	Illumina NGS	SNV, indels, fusions, CNV	10–50 ng cfDNA
FoundationOne Liquid (FoundationMedicine)	>300	Illumina NGS	SNV, indels, fusions, CNV, MSI, and blood Tumor Mutational Burden (bTMB)	20 ng cfDNA Two tubes of whole blood (8.5 mL per tube)
OncominePanCancer CFAssay (Thermo Fisher)	52	Ion GeneStudio	SNV, indels, fusions, CNV	10 mL tube of blood 5 ng input cfDNA
Guardant360 CDx (Guardant Health)	73/500 +	Guardant Health Digital Seq Platform	SNV, indels, fusions, CNV	1–2 mL plasma (5–30 ng cfDNA)
Predicine ATLAS (Predicine)	600	Illumina NGS	SNV, indels, fusions, CNV	8 mL plasma, 2 tubes of whole blood
Galleri Test (Grail)	508	Illumina NGS	SNV, indels, CNV	Plasma (single blood draw)

alteration is the V600E mutation in exon 15. BRAF mutations have been demonstrated also in a specific histotype of lung adenocarcinoma characterized by a micropapillary growth [31]. In these neoplastic forms, we demonstrated V600E and other gene mutations in exon 11 [31,32].

Among the resistance mutations induced by third-TKIs generation (osimertinib), gene MET amplification deserves consideration. In addition to conferring resistance to EGFR TKIs, this alteration promotes an aggressive phenotype and therefore a greater metastatic capacity. Another MET gene alteration involves the exon 14 skipping (METex14) and is present in 3% of lung cancers [33].

Alterations of ALK and ROS1 genes occur in 4% and 2% of NSCLC, respectively. These are translocations of a part of the gene containing kinase domain to a partner gene which is under the control of a stronger promoter. It results in a higher expression of mutated genes than the corresponding wild-type forms [34].

Nowadays to treat advanced "ALK-positive" NSCLC, the FDA and the European Medicines Agency have approved ALK TKIs, including crizotinib (first-generation), ceritinib, alectinib, brigatinib (second-generation), lorlatinib (third generation). During treatment, patients can develop resistant mutations, that are specified in Table 3.3 [35].

Emerging mutations of clinical relevance in lung cancer are rearrangements of the RET gene and KRAS mutations. RET kinase fusions are present in 1%−2% of NSCLC, caused by the rearrangement of the RET tyrosine kinase domain with different genes, most of them are KIF5B, CCDC6, and NCOA4. Thanks to the recent approval of selpercatinib, a characterization of RET fusions is now required in all patients with advanced NSCLC. Recently, DNA-based NGS is emerging as the primary screening tool for RET fusions [36].

KRAS gene mutations are present in NSCLC, especially at codons 12, 13, 59, and 61. The development of inhibitors of the mutated KRAS gene is difficult because of its high affinity to a substrate molecule GTP, so the synthetic drugs cannot efficiently compete with GTP for binding. In the last few years, many efforts have been taken for appropriate treatment. An early clinical trial with KRAS G12C inhibitor, sotorasib, produced very encouraging results and now it is approved by FDA for clinical practice, a first step forward in the treatment of lung cancer KRAS mutated tumor [37].

Malignant neoplasms of the colorectal tract are characterized by a heterogeneous and variable genetic profile and by the development over

Table 3.3 Sensitizing and resistance-inducing mutations in the ALK gene.

	Crizotinib	Ceritinib	Alectinib	Brigatinib	Lorlatinib
Sensitizing mutations	L1198F	L1196M, I1171T, G1269A, S1206Y,	G1269A, F1174L S1206Y, L1152R 1151Tins	L1152R, F1174C G1269A, S1206Y, F1245C, I1171T, D1203N, E1210K, 1151Tins,	G1202R, G1269A L1196M,
Resistance-inducing mutations	L1196Q, S1206Y L1198P, F1174L L1196M, G1269A C1156Y, I1171T/N G1202R, D1203N 1151Tins, L1152R	D1203, G1202R R1275Q, L1152P/R L1198F, C1156Y/T F1174C/V	G1202R I1171T, V1180L	G1202R, S1206C/F F1174V + L1198F	L1198F

time of additional mutations correlated with tumor progression. Massive sequencing of circulating cfDNA can allow following the dynamic evolution of the disease, directing therapy in a timely manner [38].

Like for other neoplasms, through cftDNA analysis it is possible to reveal genomic driver colorectal alterations, (e.g., RAS, BRAF, ERBB2, and MET), as well as other cancer-related genes associated with resistance to anti-EGFR therapy. By monitoring cftDNA under anti-EGFR therapy (cetuximab, panitumumab), the emergence of genomic alterations can be detected, as acquired resistance mechanisms in specific genes, including MET, ERBB2, FLT3, EGFR, and MEK. Analysis of cftDNA can also identify predictive biomarkers to immune checkpoint inhibitors, such as mutations in mismatch repair genes, microsatellite instability, and TMB [39,40].

Survival of breast cancer (BC) patients has significantly improved thanks to the spread of screening programs, which help in prevention and early cancer detection. Furthermore, the increase in knowledge about the molecular characterization of this pathology represents a guide in the development of new and even more personalized diagnostic, prognostic, and therapeutic tools.

HER2 is an overexpressed biomarker from 10% to 25% of BCs and causes an aggressive phenotype with higher metastatic potential and a worse prognosis due to stimulation of the MAPK and PI3K/AKT signaling pathway [1]. HER2 amplification is considered a prognostic indicator and is currently being re-evaluated in new metastatic lesions.

The HER2-specific treatments are improving patient survival. The assessment of the molecular status of the HER2 gene by NGS analysis on circulating cfDNA shows concordance rates up to 90% with the HER2 status analyzed on tissue biopsy. Therefore, NGS evaluation is a feasible strategy for monitoring response to therapy carried out primarily with the monoclonal antibody trastuzumab [41].

Although HER2 overexpression/amplification appears to be required for a response to anti-HER2 therapy, many patients (30%) with aberrant HER2 status do not respond to monoclonal antibody therapy, and responding patients develop resistance over time. This implies that new gene alterations develop which can negatively affect the response to anti-HER2 therapies.

In the context of acquired mutations, the gene alterations of PIK3CA frequent in some histotypes of BC are of particular interest [42].

Several studies conducted on BC have highlighted the clinical utility of liquid biopsy which emerged as an interesting new diagnostic tool, as also stated for other tumors [43,44].

The frequency of germline BRCA mutations in unselected BC patients is around 5% but can be as high as 16%—40% in triple-negative breast cancer (TNBC) patients. Liquid biopsy has recently been used to monitor the clinical response to PARP inhibitors treatment in TNBC with BRCA1 and 2 mutations [45,46].

BRCA1 and 2 alterations affect DNA double-strand breaks repair due to deficient homologous recombination repair mechanism. When BRCA 1—2 are compromised as well as the function of PARP enzymes, that inhibit DNA single-strand break repair, DNA replication errors multiply, inducing cancer cell death, a mechanism known as synthetic lethality [47,48].

The use of PARP inhibitors in germline BRCA mutated BCs has been shown to improve patient outcomes, regardless of tumor HR/HER2 status. However, resistance mechanisms to PARP inhibitors can occur, for example, secondary intragenic deletions or reverse mutations. cftDNA may help predict resistance to PARP inhibitors with the detection of these resistant mutations [49].

Some evidence highlighted that circulating cfDNA analysis could represent a non-invasive and feasible approach for early diagnosis of tumor progression and for monitoring the therapeutic response in patients affected by Melanoma, a neoplastic disease with an increasing worldwide prevalence and high mortality. Mutations in the BRAF, NRAS, or KIT genes are present in over 60% of melanoma cases [50]. The presence of BRAF mutation is necessary for the combined BRAF and MEK-specific inhibitors. In this case, after initial treatment, the monitoring shows a rapid and marked decrease of the BRAF V600E circulating mutated alleles. This decrease is related to response to treatment, as shown by the radiological assessment [6]. On the other hand, an increase of mutated alleles, at progression, was observed and in a significant fraction of patients, it precedes the clinical progression [51].

In summary, multigenic analysis on circulating cfDNA, based on MPS, could significantly increase the information on tumor heterogeneity and let get evidence on a broad spectrum of the genes involved in the mechanisms of resistance and progression. Although the recent data of literature are to be considered strongly suggestive of the possible use of this therapeutic strategy in clinical practice for some neoplastic forms, further validations are needed on a larger number of patients for other cancers.

3.4 Expert opinion

Nowadays, several prospective clinical trials in oncological patients incorporate longitudinal ctDNA genotyping to monitor clonal changes and guide treatment decisions. These clinical studies are fundamental to assess the clinical applicability of ctDNA in metastatic and localized solid tumors, whereas results of ongoing clinical studies of liquid biopsy as a tool for screening and detection of premalignant disease are highly awaited.

3.5 Key points

- Careful profiling of the tumor molecular landscape is mandatory to tailor treatment decisions.
- Have reached a better understanding of the particular molecular events accountable for tumor progression and onset.

3.6 Hints for deeper insight

- Recent progress in biotechnology (i.e., next-generation sequencing, digital PCR) has led to the development of liquid biopsy. With increasing modern technologies, liquid biopsy could currently be used in an easy and reproducible way in daily clinical practice.
- Liquid biopsy applications combine molecular diagnosis, determination of tumor load as a surrogate marker of early response to oncological treatment, monitoring mutations of resistance to targeted therapy, and detection of minimal residual disease after cancer resection.

References

[1] Fernández-Lázaro D, Hernández JLG, García AC, Castillo ACD, Hueso MV, Cruz-Hernández JJ. Clinical perspective and translational oncology of liquid biopsy. Diagnostics (Basel) 2020;10(7):443.

[2] Hugenschmidt H, Labori KJ, Borgen E, Brunborg C, Schirmer CB, Seeberg LT, et al. Preoperative CTC-detection by CellSearch® is associated with early distant metastasis and impaired survival in resected pancreatic cancer. Cancers (Basel) 2021;13 (3):485.

[3] Andree KC, van Dalum G, Terstappen LW. Challenges in circulating tumor cell detection by the CellSearch system. Mol Oncol 2016;10(3):395—407.

[4] Galvis MM, Romero CS, Bueno TO, Teng Y. Toward a new era for the management of circulating tumor cells. Adv Exp Med Biol 2021;1286:125—34.

[5] Gašperšič J, Videtič, Paska A. Potential of modern circulating cell-free DNA diagnostic tools for detection of specific tumour cells in clinical practice. Biochem Med (Zagreb) 2020;30(3):030504.

[6] Kamińska P, Buszka K, Zabel M, Nowicki M, Alix-Panabières C, Budna-Tukan J. Liquid biopsy in melanoma: significance in diagnostics, prediction and treatment monitoring. Int J Mol Sci 2021;22(18):9714.

[7] Tsui DWY, Blumenthal GM, Philip R, Barrett JC, Montagut C, Bramlett K, et al. Development, validation, and regulatory considerations for a liquid biopsy test. Clin Chem 2020;66(3):408—14.

[8] News ROCHE, FDA grants first liquid biopsy approval to the Roche cobas® EGFR mutation test v2, Pleasanton, June 1; 2016.

[9] Mohammad Kazem Aghamir S, Heshmat R, Ebrahimi M, et al. Liquid biopsy: the unique test for chasing the genetics of solid tumors. Epigenet Insights 2020;13. 2516865720904052.

[10] Lynch TJ, Bell DW, Sordella R, et al. Activating mutations in the epidermal growth factor receptor underlying responsiveness of nonsmall-cell lung cancer to gefitinib. N Engl J Med 2004;350:2129—39 25.

[11] Paez JG, Janne PA, Lee JC, et al. EGFR mutations in lung cancer: correlation with clinical response to gefitinib therapy. Science 2004;304:1497—500.

[12] Daniel Stetson MS, Ambar Ahmed MS. Orthogonal comparison of four plasma NGS tests with tumor suggests technical factors are a major source of assay discordance. JCO Precis Oncol 2019;3:1—9.

[13] Cheng F, Su L, Qian C. Circulating tumor DNA: a promising biomarker in the liquid biopsy of cancer. Oncotarget. 2016;7(30):48832—41.

[14] Bettegowda C, Sausen M, Leary RJ, Kinde I, Wang Y, Agrawal N, et al. Detection of circulating tumor DNA in early- and late-stage human malignancies. Sci Transl Med 2014;6(224): 224ra24.

[15] Sha D, Jin Z, Budczies J, Kluck K, Stenzinger A, Sinicrope FA. Tumor mutational burden as a predictive biomarker in solid tumors. Cancer Discov 2020;10 (12):1808—25.

[16] Gu W, Miller S, Chiu CY. Clinical metagenomic next-generation sequencing for pathogen detection. Annu Rev Pathol 2019;14:319—38.

[17] Kamps R, Brandão RD, Bosch BJ, Paulussen AD, Xanthoulea S, Blok MJ, et al. Next-generation sequencing in oncology: genetic diagnosis, risk prediction and cancer classification. Int J Mol Sci 2017;18(2):308.

[18] Zhong Y, Xu F, Wu J, Schubert J, Li MM. Application of next generation sequencing in laboratory medicine. Ann Lab Med 2021;41(1):25—43.

[19] Foox J, Tighe SW, Nicolet CM, Zook JM, Byrska-Bishop M, Clarke WE, et al. Performance assessment of DNA sequencing platforms in the ABRF next-generation sequencing study. Nat Biotechnol 2021;39(9):1129—40.

[20] Bruno R, Fontanini G. Next generation sequencing for gene fusion analysis in lung cancer: a literature review. Diagnostics (Basel) 2020;10(8):521.

[21] Newman AM, Bratman SV. To J, et al. An ultrasensitive method for quantitating circulating tumor DNA with broad patient coverage. Nat Med 2014;20(5):548—54.

[22] Newman AM, Lovejoy AF, Klass DM, et al. Integrated digital error suppression for improved detection of circulating tumor DNA. Nat Biotechnol 2016;34(5):547—55.

[23] Honoré N, Galot R, van Marcke C, Limaye N, Machiels JP. Liquid biopsy to detect minimal residual disease: methodology and impact. Cancers (Basel) 2021;13 (21):5364.

[24] Phallen J, Sausen M, Adleff V, Leal A, Hruban C, White J, et al. Direct detection of early-stage cancers using circulating tumor DNA. Sci Transl Med 2017;9(403): eaan2415.

[25] Marchetti A, Martella C, Felicioni L, Barassi F, Salvatore S, Chella A, et al. EGFR mutations in non-small-cell lung cancer: analysis of a large series of cases and development of a rapid and sensitive method for diagnostic screening with potential implications on pharmacologic treatment. J Clin Oncol 2005;23(4):857—65.

[26] Marchetti A, Palma JF, Felicioni L, Buttitta F, et al. Early prediction of response to tyrosine kinase inhibitors by quantification of EGFR mutations in plasma of NSCLC patients. J Thorac Oncol 2015;10(10):1437—43.

[27] Del Re M, Rofi E, Cappelli C, Puppo G, Crucitta S, Valeggi S, et al. The increase in activating EGFR mutation in plasma is an early biomarker to monitor response to osimertinib: a case report. BMC Cancer 2019;19(1):410.

[28] Papadopoulou E, Tsoulos N, Tsantikidi K, Metaxa-Mariatou V, Stamou PE, Kladi-Skandali A, et al. Clinical feasibility of NGS liquid biopsy analysis in NSCLC patients. PLoS One 2019;14(12):e0226853.

[29] Zulato E, Tosello V, Nardo G, Bonanno L, Del Bianco P, Indraccolo S. Implementation of next generation sequencing-based liquid biopsy for clinical molecular diagnostics in non-small cell lung cancer (NSCLC) patients. Diagnostics (Basel) 2021;11(8):1468.

[30] Buttitta F, Felicioni L, Marchetti A, Lorito AD, et al. Early prediction of resistance to tyrosine kinase inhibitors by plasma monitoring of EGFR mutations in NSCLC: a new algorithm for patient selection and personalized treatment. Oncotarget. 2020;11 (11):982—91.

[31] Marchetti A, Felicioni L, Malatesta S, Grazia Sciarrotta M, Guetti L, Chella A, et al. Clinical features and outcome of patients with non-small-cell lung cancer harboring BRAF mutations. J Clin Oncol 2011;29(26):3574—9.

[32] Fernandes MGO, Sousa C, Pereira Reis J, Cruz-Martins N, et al. Liquid biopsy for disease monitoring in non-small cell lung cancer: the link between biology and the clinic. Cells. 2021;10(8):1912.

[33] Fernandes M, Jamme P, Cortot AB, Kherrouche Z, Tulasne D. When the MET receptor kicks in to resist targeted therapies. Oncogene. 2021;40(24):4061—78.

[34] Pan Y, Deng C, Qiu Z, Cao C, Wu F. The resistance mechanisms and treatment strategies for ALK-rearranged non-small cell lung cancer. Front Oncol 2021;11:713530.

[35] Sharma GG, Mota I, Mologni L, Patrucco E, Gambacorti-Passerini C, Chiarle R. Tumor resistance against ALK targeted therapy-where it comes from and where it goes. Cancers (Basel) 2018;10(3):62.

[36] Yang SR, Schultheis AM, Yu H, Mandelker D, Ladanyi M, Büttner R. Precision medicine in non-small cell lung cancer: current applications and future directions. Semin Cancer Biol 2020;S1044-579X(20) 30164-30164.

[37] Imyanitov EN, Iyevleva AG, Levchenko EV. Molecular testing and targeted therapy for non-small cell lung cancer: current status and perspectives. Crit Rev Oncol Hematol 2021;157:103194.

[38] Heidrich I, Ačkar L, Mossahebi Mohammadi P, Pantel K. Liquid biopsies: potential and challenges. Int J Cancer 2021;148(3):528—45.

[39] Erve I, Greuter MJE, Bolhuis K, Vessies DCL, Leal A, Vink GR, et al. Diagnostic strategies toward clinical implementation of liquid biopsy RAS/BRAF circulating tumor DNA analyses in patients with metastatic colorectal cancer. J Mol Diagn 2020;22(12):1430—7.

[40] Naidoo M, Piercey O, Tie J. Circulating tumour DNA and colorectal cancer: the next revolutionary biomarker? Curr Oncol Rep 2021;23(12):140.

[41] Li J, Guan X, Fan Z, Ching LM, Li Y, Wang X, et al. Non-invasive biomarkers for early detection of breast cancer. Cancers (Basel) 2020;12(10):2767.

[42] Buttitta F, Felicioni L, Barassi F, Martella C, Paolizzi D, Fresu G, et al. PIK3CA mutation and histological type in breast carcinoma: high frequency of mutations in lobular carcinoma. J Pathol 2006;208(3):350−5.

[43] Keup C, Benyaa K, Hauch S, Sprenger-Haussels M, Tewes M, Mach P, et al. Targeted deep sequencing revealed variants in cell-free DNA of hormone receptor-positive metastatic breast cancer patients. Cell Mol Life Sci 2020;77(3):497−509.

[44] Tay TKY, Tan PH. Liquid biopsy in breast cancer: a focused review. Arch Pathol Lab Med 2021;145(6):678−86.

[45] Malone KE, Daling JR, Doody DR, Hsu L, Bernstein L, Coates RJ, et al. Prevalence and predictors of BRCA1 and BRCA2 mutations in a population-based study of breast cancer in white and black American women ages 35 to 64 years. Cancer Res 2006;66(16):8297−308.

[46] Fostira F, Tsitlaidou M, Papadimitriou C, Pertesi M, Timotheadou E, Stavropoulou AV, et al. Prevalence of BRCA1 mutations among 403 women with triple-negative breast cancer: implications for genetic screening selection criteria: a Hellenic Cooperative Oncology Group Study. Breast Cancer Res Treat 2012;134(1):353−62.

[47] Ashworth A. A synthetic lethal therapeutic approach: poly(ADP) ribose polymerase inhibitors for the treatment of cancers deficient in DNA double-strand break repair. J Clin Oncol 2008;26(22):3785−90.

[48] Dedes KJ, Wilkerson PM, Wetterskog D, Weigelt B, Ashworth A, Reis-Filho JS. Synthetic lethality of PARP inhibition in cancers lacking BRCA1 and BRCA2 mutations. Cell Cycle 2011;10(8):1192−9.

[49] Chan JCH, Chow JCH, Ho CHM, Tsui TYM, Cho WC. Clinical application of circulating tumor DNA in breast cancer. J Cancer Res Clin Oncol 2021;147 (5):1431−42.

[50] Pinzani P, D'Argenio V, Del Re M, Pellegrini C, Cucchiara F, Salvianti F, et al. Updates on liquid biopsy: current trends and future perspectives for clinical application in solid tumors. Clin Chem Lab Med 2021;59(7):1181−200.

[51] Sacco A, Forgione L, Carotenuto M, Luca A, Ascierto PA, Botti G, et al. Circulating tumor DNA testing opens new perspectives in melanoma management. Cancers (Basel) 2020;12(10):2914.

Further reading

Diaz and Bardelli, 2014Diaz LA, Bardelli A. Liquid biopsies: genotyping circulatingntumor DNA. J Clin Oncol 2014;32:579−86.

Rolfo et al., 2020Rolfo C, Cardona AF, et al. Challenges and opportunities of cfDNA analysis implementation in clinical practice: perspective of the International Society of Liquid Biopsy (ISLB). Crit Rev Onc 2020;102:978.

CHAPTER 4

Current clinically validated applications of liquid biopsy

E. Capoluongo[1,2], C. Rolfo[3], A. Galvano[4], V. Gristina[4], A. Perez[4], N. Barraco[4], M. La Mantia[4], L. Incorvaia[4], G. Badalamenti[4], A. Russo[4,*] and V. Bazan[5,*]

[1]Department of Molecular Medicine and Medical Biotechnology, University of Naples Federico II, Naples, Italy
[2]CEINGE, Advances Biotecnologies, Naples, Italy
[3]Center for Thoracic Oncology, Tisch Cancer Institute, Mount Sinai Medical System & Icahn School of Medicine, Mount Sinai, NY, United States
[4]Department of Surgical, Oncological, and Oral Sciences, University of Palermo, Palermo, Italy
[5]Department of Biomedicine, Neuroscience and Advanced Diagnostics (Bi.N.D.), University of Palermo, Palermo, Italy

Learning objectives

By the end of the chapter, the reader will:

- Learn the applications of liquid biopsy in non-small cell lung cancer (NSCLC) clinical practice
- Have a deep knowledge of the emerging clinical application of liquid biopsy in different tumors

4.1 Circulating tumor DNA in advanced non-small cell lung cancer

In current clinical practice, LB is mainly used to analyze *EGFR* gene mutational status in advanced NSCLC patients [1]. Until a few years ago, the main source of biological material to perform molecular testing was tissue. However, in advanced NSCLC tissue may frequently not be suitable for molecular testing, especially when considering the emerging need to test an increasing number of prognostic and predictive biomarkers in this specific setting. Molecular testing in advanced NSCLC is demanding and must be contextualized within the natural history of the disease meaning that it may be required to obtain genetic information at diagnosis, during treatment, and at progression. Although so far, re-biopsy was

---

* Russo A. and Bazan V. should be considered equally colast authors.

Liquid Biopsy
DOI: https://doi.org/10.1016/B978-0-12-822703-9.00010-7
© 2023 Elsevier Inc.
All rights reserved.

the only choice for clinicians, data emerging from several clinical studies and metanalysis have proved that LB could be a new weapon in clinicians' hands to obtain a real-time portrait of the tumor in advanced NSCLC [2−4]. These shreds of evidence have led to the first clinical indication of LB in clinical practice. *EGFR* mutational status can be evaluated from plasma ctDNA when (1) the cyto-histological sample does not contain sufficient (both in terms of quantity and quality) viable neoplastic cells for molecular testing; or (2) when patients' general conditions do not allow the execution of a tissue biopsy. Considering the proven high diagnostic concordance between ctDNA and tissue, LB can be used in two clinical scenarios: (1) at diagnosis in treatment-naïve advanced NSCLC patients whenever there is insufficient or poor quality cyto-histological material to perform molecular testing; (2) in *EGFR*-mutant patients progressing on first-line therapy with first or second-generation TKI (tyrosine kinase inhibitors) to detect resistance mutation p.T790M in *EGFR* exon 20 as a predictive biomarker of response to third-generation TKI [5].

Despite *EGFR* mutations, genetic alterations in other oncogenic driver genes have been shown to influence targeted treatment sensitivity and/or resistance. A point mutation in the *BRAF* gene, rearrangements of *ALK*, *ROS1*, and *NTRK* genes as well as MET exon 14 skipping, MET and HER2 amplification, and the p.G12C mutation in *KRAS* genes are all involved in positively or negatively defining treatment response. Nevertheless, despite being feasible, none of the aforementioned alterations could be tested on LB in clinical practice but pertaining only to clinical studies [6]. However, there is an emerging body of evidence indicating that targeted treatments are effective also when genetic alterations are detected on ctDNA instead of tissue, thus suggesting that LB could be used besides the actual clinical indications in selected cases with specific unmet medical needs that should be always evaluated within a tumor molecular board.

4.1.1 Treatment-naïve advanced non-small cell lung cancer patients

In this setting, ctDNA analysis represents a valuable alternative whenever tissue samples are not suitable for detecting *EGFR* mutational status (Fig. 4.1). Sample suitability depends on both quantities, meaning the percentage of tumor cells among normal cells, and quality which takes into account pre-analytical phases (e.g., fixation and inclusion) that could profoundly impair the property of extracted DNA which in turn lead to unsuccessful molecular testing. Using a simple blood withdrawal, it might

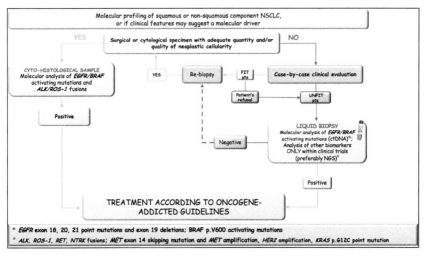

Figure 4.1 Diagnostic algorithm for treatment-naive advanced oncogene-addicted NSCLS.

be possible to overcome this tissue-related issue; however, it must be underlined that whenever the analysis on LB is negative (absence of *EGFR* activating mutations) re-biopsy should be preferred to avoid the risk of a false-negative result. Indeed, ctDNA concentration is variable and may be below the test limit of detection (LOD), thus incurring the risk of missing the alteration.

Technically, ctDNA analysis can be conducted using any kind of method that enables the detection of genetic alteration (e.g., point mutations, small insertion, and deletion). Therefore, any sequencing-based or probe-based approach is suitable for this analysis, however, what is not irrelevant is the technique sensitivity. Real-Time Polymerase Chain Reaction (RT-PCR) is currently the gold standard method for the analysis of ctDNA; this method is considerably diffused in molecular diagnostic laboratories as it assures proper sensitivity and turn-around time (TAT) at low costs. RT-PCR relies on the use of a probe that can be modified to improve diagnostic sensitivity (e.g., the Amplification Refractory Mutation System—ARMS/SCORPION), reaching a LOD of 0.5%. Therefore, this method is suitable to detect low percentages of mutated alleles among high quantities of wild-type genomic DNA as in the case of ctDNA. The digital polymerase chain reaction (dPCR) represents a technological advancement of the classic RT-PCR [7]; this innovative approach is intended to transform the exponential, analog nature of the

PCR into a linear, digital (or binary) signal. There are different dPCR platforms (droplet digital PCR—ddPCR; solid digital PCR—sdPCR, and BEAMing—Beam, Emulsion, Amplification, Magnetics) but they are all based on the sample principle; single DNA molecules are spread out inside the bioreactors (droplets or wells) according to Poisson distribution [7]. Each bioreactor contains the master mix and specific fluorescent probes for both wild-type and mutated alleles. After the PCR amplification steps, the analysis is based on fluorescence detection and the results are obtained by counting the number of *positive* and *negative* reactions. A reaction is defined as *positive* when the fluorescence of the specific probe is detected; accordingly, a *negative* reaction is defined by the absence of the specific probe fluorescence emission. By partitioning the amplification reactions, it is possible to obtain both qualitative information as well as absolute quantification of even a small number of mutated alleles in a background of wild-type alleles. Therefore, dPCR can reach greater sensitivity (0.1%– 0.01%) compared to Real-Time PCR [8]. Both RT-PCR and dPCR methods are probe-based approaches that have one main limitation compared to the sequencing-based approach. Indeed probe-based methods offer the possibility to only detect known mutations thus impairing the possibility to identify very rare mutations or new alterations. Conversely, a sequencing-based method such as NGS (especially targeted sequencing applications) offers the great opportunity to investigate in a single experiment multiple genes and multiple known and unknown alterations. Despite, initially, NGS sensitivity in the context of LB was poor (around 1%–2%) [9] with the development of new and more sensible applications it is now possible to reach a sensitivity below 1% (0.1%–0.01%), that perfectly fits with the needs of ctDNA testing [10].

In treatment-naïve advanced NSCLC patients, NGS should be preferred over other traditional technologies considering its greater sensitivity and specificity, but mostly because it allows the detection of less common but potentially targetable alterations [11]. However, laboratories working with NGS must be adequately trained and therefore this method is still not widely used in the context of LB [12].

4.1.2 Advanced non-small cell lung cancer patients progressing during tyrosine kinase inhibitors

All *EGFR*-mutant patients that received 1st or 2nd generation TKIs (gefitinib, erlotinib, or afatinib) as first-line therapy, experiencing disease progression, must be tested for the mutation p.T790M arising in *EGFR* exon

20. Indeed, this mutation represents the main resistance mechanism to 1st or 2nd generation TKIs and it determines first-line treatment failure in approximately 50%–60% of patients [1]. Clearly, advanced NSCLC patients at this stage are even less fit for a re-biopsy. Thus, considering the high diagnostic accuracy of ctDNA analysis it is reasonable to use LB as an upfront strategy to detect p.T790M mutation (preferably using dPCR or RT-PCR) [1].

Based on a positive result obtained in LB (meaning the presence of p. T790M mutation) clinicians can switch from first/second generation TKI to third-generation TKI (osimertinib). However, in case of a negative result (no detection of p.T790M mutation), whenever the patient's condition allows, it should be always proposed a re-biopsy. More recently, in the FLAURA trial osimertinib showed superior efficacy to that of standard *EGFR*-TKIs in the first-line treatment of *EGFR*-mutant advanced NSCLC, with a similar safety profile and lower rates of serious adverse events. Therefore, osimertinib represents the best therapeutic option in treatment-naïve *EGFR*-mutant patients. Consequently, the identification of resistance mutation upon progression on third-generation TKI is a secondary option. Indeed, once the possibility of any further use of targeted therapies is exhausted, *EGFR*-positive patients are candidates for chemotherapy according to the schemes used in nonsquamous histology. Nevertheless, as previously mentioned, a LB may still be useful for detecting other resistance mechanisms (e.g., *MET* and *HER2* amplification, *ALK* mutations, other *EGFR* mutations) to tailor targeted treatment in the context of clinical studies (Fig. 4.2).

4.2 Emerging clinical applications of liquid biopsy

The development of NGS technologies for the study of ctDNA has greatly broadened the possibilities of the clinical application of LB. The availability of large gene panels allows obtaining a complete genomic profile of the neoplasm even starting from a few nanograms of ctDNA usually arising from peripheral blood plasma extraction. Several studies have shown that the genetic profile derived from the ctDNA analysis is similar to that obtained from the analysis of tissue tumor DNA, especially in lung cancer, even if there are now applications to almost all neoplasms [13–16]. In particular, the possibility of identifying different types of genetic alterations with NGS, also by analyzing ctDNA, has been highlighted in various publications, with response rates to target

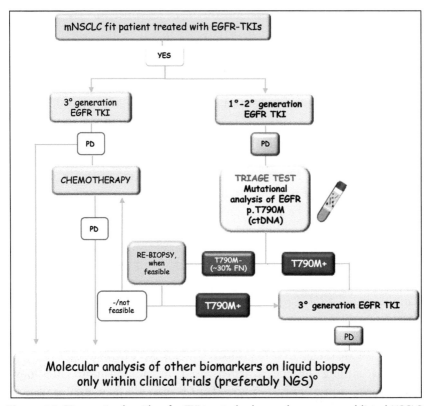

Figure 4.2 Diagnostic algorithm for TKI treated advanced oncogene-addicted NSCLS.

treatments similar to those detected for genetic alterations identified by tumor tissue analysis [17]. Therefore, in the light of this scientific evidence, the analysis of ctDNA with NGS may represent an alternative approach for the identification of driver mutations in human neoplasms as well as complex markers such as microsatellite instability (MSI) or tumor mutational burden. However, at moment the use of LB should only be considered in the event of nonavailability of tumor tissue or in specific clinical settings. In fact, the success chances of LB analysis are related to the amount of ctDNA present in the peripheral blood, which can significantly limit the sensitivity of the test. A further limit to the use of this approach may also be represented by the absence of specific studies and clinical trials on LB. In the case of gene fusions, the technical limitations related to DNA analysis must also be evaluated. Especially for complex fusions, such as those of NTRK (Neurotrophic Receptor Tyrosine

Kinase) characterized by large intronic regions or Fibroblast Growth Factor Receptor which have multiple partners, the possibilities of identification by ctDNA analysis are limited and, therefore, this analysis could give false-negative results. The possibility of identifying gene fusions is generally greater by analyzing RNA, however, its reduced stability often limits the chances of success. Furthermore, it must be emphasized that each NGS panel has its own sensitivity and specificity that can vary significantly. Therefore, the results obtained with a specific panel cannot be generalized. Each panel must undergo a necessary validation and/or verification phase to determine its reliability and its limits. Finally, the possibility of analyzing large genomic regions with NGS has greatly expanded the use of LB for the study of resistance mechanisms to biological drug regimens [18−20]. Indeed, thanks to the NGS analysis, several mechanisms of resistance to new-generation TKIs have been discovered in lung cancer as well as *BRCA* reversal mutations in ovarian cancer patients treated with Poly ADP-ribose polymerase inhibitors. These approaches must be considered experimental in some way, as specific therapies for these alterations are not currently available in clinical practice. It is undeniable, however, that the acquisition of this information can facilitate treatments personalization and increase patients' enrollment in clinical trials. Therefore, although not indicated in clinical practice, the use of NGS testing on LB to study resistance mechanisms in patients treated with biological drugs can certainly provide relevant information in academic centers that have many available clinical trials. Finally, given data complexity arising from large gene panels and the need for their correct interpretation in the context of a specific clinical scenario, it is recommended that the results of NGS tests are discussed in Molecular Tumor Boards.

4.2.1 Colorectal cancer

The applications of LB for the detection of ctDNA in colorectal cancer (CRC) are an emerging field of research and mainly concern (Fig. 4.3):
- the early stage of disease for prognostic evaluation and to tailor the choice of adjuvant therapy;
- the advanced stage of disease for the analysis of *RAS* and *BRAF* genetic alterations and the monitoring of molecular targeted therapies.

Stage I-III disease—The ability to use ctDNA as a marker of minimal residual disease (MRD) has become an emerging area of clinical research for localized CRC. In this context, a correlation has already been

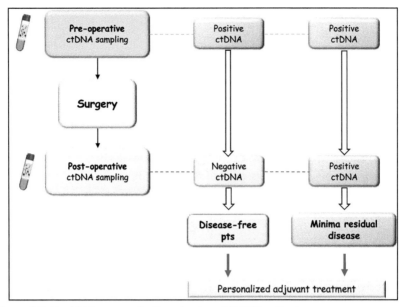

Figure 4.3 Liquid biopsy applications in stage I-III CRC patients.

observed between the presence of ctDNA after surgical resection of the primary tumor and the recurrence of disease, both at stages II and III [21−23]. Moreover, the evaluation of more than one variant as well as the use of serial sampling increases the accuracy in predicting the presence of MRD [23]. Unfortunately, the data currently available derive from the heterogeneous cohort, very often with a limited number of patients, not always treated with an adjuvant regimen and there is no contextualization with respect to other clinical, immunological, and molecular markers allowing to highlight the importance of this approach with respect to other standards or emerging biomarkers (e.g., Immunoscore, CDX-2, MSI). Therefore, the application of LB for early-stage disease is still experimental.

Stage IV disease—Numerous studies have demonstrated the feasibility of performing *RAS* liquid testing as a potential substitute for tumor tissue analysis in metastatic CRC. The concordance between the two approaches with current sequencing techniques varies from 60% to 80% [24], although it must be considered that the tumor and peripheral blood are two distinct tissues and therefore it is not reasonable to expect a perfect concordance. Moreover, the discrepancies observed in terms of specificity—assuming the tumor tissue as a reference—are justified by the fact that LB can overcome spatial and temporal heterogeneity that usually limits tissue analysis. Certainly, LB offers the advantages of a relatively noninvasive approach as well as more flexibility, both for allowing the easy determination of the mutational status at the exact time of therapeutic intervention with an anti-EGFR, and for the reduced TAT. However, there is insufficient clinical evidence to establish the cut-off of mutated *RAS* alleles determined on peripheral blood that confers resistance to anti-EGFR therapy. Therefore, ctDNA analysis should be considered at the moment only within specific clinical trials (Fig. 4.4).

4.2.2 Breast cancer

Breast cancer is characterized by molecular heterogeneity given the presence of cellular clones carrying different gene alterations guiding their growth and proliferation. The presence of molecular alterations determines, in some cases, tumor responsiveness to targeted treatment, but often the co-presence of a small percentage of cellular clones carrying other mutations leads to therapy failure [25]. Several studies have shown the clinical utility of LB in breast cancer patients in identifying biomarkers predicting response or resistance to treatment, as well as quantitative monitoring of ctDNA during treatment [26−30]. In particular, LB analysis of frequently mutated genes in breast cancer (i.e., Estrogen Receptor 1

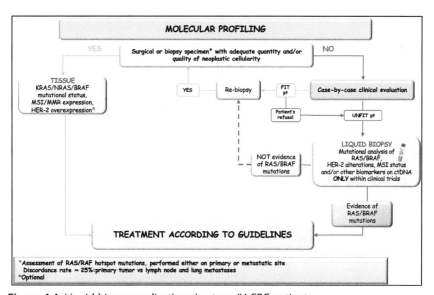

Figure 4.4 Liquid biopsy applications in stage IV CRC patients.

[ESR1], PhosphoInositide 3-kinase [PI3K], tumor protein p53 [p53]) have also been matched with the evaluation of tumor burden, the identification and monitoring of residual disease in patients undergoing curative surgery [31], and as a significant prognostic factor [32] (Fig. 4.5). It has recently been shown that mutations in the *ESR1* gene have a predictive role in resistance to treatment with CDK4/6 inhibitors. *ESR1* mutations were analyzed in the ctDNA of 1017 patients with metastatic breast cancer before and after 1-month treatment with palbociclib in combination with antiaromatase as a first-line approach. Interestingly, the presence of the mutations significantly reduced progression-free survival (PFS) (*ESR1* wild type vs *ESR1* mut 26.7 vs 11 months, p <0.001). Furthermore, in the group of mutated patients (3.2%), it was also shown that the clearance of *ESR1* mutations in ctDNA after 1 month of treatment predicts a longer survival, compared to the group of patients who maintained detectable ctDNA mutation amounts (*ESR1* mut cleared vs *ESR1* mut detected 24.1 vs 7.4 months, P <0.001). We need to wait for the results from the PADA-1 trial to verify whether the screening for *ESR1* gene mutations is clinically relevant.

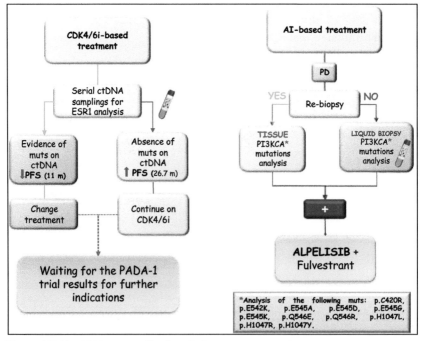

Figure 4.5 Liquid biopsy applications in breast cancer patients.

Treatment with alpelisib, a *PIK3CA* gene inhibitor, was approved by the FDA in 2019 for *PIK3CA*-mutated metastatic breast cancer patients. The study highlighted that treatment with alpelisib prolonged PFS in patients who received a previous line of endocrine therapy [33]. Some studies have shown that the onset of mutations in the *PIK3CA* gene is one of the mechanisms of acquired resistance to hormone therapy [34,35]. For this reason, alpelisib treatment has been approved by the Food and Drug Administration based on the presence of *PIK3CA* mutations on both tissues (if available) and LB. Treatment with the PI3K inhibitor is approved on the basis of the presence of one of the following mutations in the *PIK3CA* gene: p.C420R, p.E542K, p.E545A, p.E545D, p.E545G, p.E545K, p.Q546E, p.Q546R, p.H1047L, p.H1047R, p.H1047Y. Although technological approaches such as NGS may allow a better evaluation of the clonal heterogeneity of the disease, for the identification of multiple biomarkers and an evaluation of the neoplastic molecular evolution during the clinical follow-up, these remain, to date, recommended only within clinical studies. In conclusion, regarding breast cancer, it is likely that the plasma analysis of the *PIK3CA* gene mutations will soon be recommended in clinical practice.

4.2.3 Melanoma

Several studies have shown a possible clinical utility of LB in melanoma patients, both in the identification of *BRAF* and *NRAS* mutations to set up the treatment (if the tissue is not available) and for the quantitative monitoring of the ctDNA during treatment [36−40]. In particular, the analysis of mutations in *BRAF* and *NRAS* genes in ctDNA using RT-PCR or dPCR methods has been associated with the evaluation of tumor burden [41], for MRD evaluation in patients undergoing radical surgery [42]. Moreover, the identification of *BRAF* and *NRAS* mutation in ctDNA is a significant prognostic factor in patients with stage II/III melanoma [43,44] or with metastatic disease [41]. In addition, ctDNA analysis has been proposed as a useful biomarker of response to therapy with kinase inhibitors or immunotherapy, and of the early appearance of resistance to treatment [45−48]. As happened in other neoplastic contexts, also for melanoma, NGS multigenic panels have been recently introduced for the study of ctDNA allowing thus a wide extension of the analysis also to cases that do not harbor mutations in *BRAF/NRAS* [49]. Furthermore, this method allows a better study of clonal heterogeneity in metastatic disease and a noninvasive evaluation of the neoplastic molecular evolution during clinical follow-up. In addition to the study of ctDNA, other biomarkers in LB have been proposed in patients with melanoma. Among these, evaluation of the exosomal expression of PD-L1 (Programmed Death-Ligand 1) was found to be a predictive marker of response to immunotherapy [50]. Interestingly, the concordance between ctDNA and tissue increases proportionally to the stage of the disease, rising from 25% to 40% in stages II/III, and up to 70% in stages IV [43,44,51]. In summary, the use of ctDNA in metastatic melanoma has proved to be a useful tool for identifying both resistance and predictive biomarkers to targeted therapy but also for monitoring treatment response during the course of targeted therapy and immunotherapy [46,51,52]. However, given the amount of scientific evidence reported in the literature to support the genetic testing in the ctDNA, in selected cases discussed within multidisciplinary groups, the use could be recommended also outside of clinical trials (Fig. 4.6).

4.3 Liquid biopsy application in clinical research

4.3.1 Monitoring of treatment response in lung cancer

The use of ctDNA is increasingly proving its usefulness in monitoring treatment response in several solid tumor patients with therapeutically

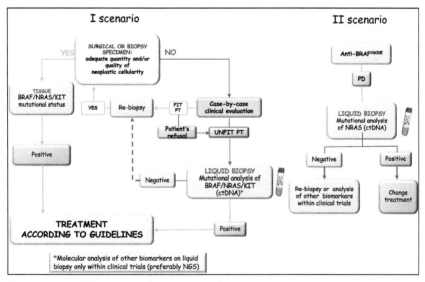

Figure 4.6 Liquid biopsy applications in melanoma patients.

targetable alterations. The traditional radiological techniques do not produce the same type of information and this can have consequences on the ability to understand the molecular mechanisms of resistance by limiting the diagnostic and therapeutic potentialities. The identification of MRD after surgery, as well as the ability to predict recurrence, remains an important medical unmet need, to which ctDNA evaluation can give important answers. ctDNA has a relatively high clearance, making it an excellent biomarker reflecting in real-time the adaptive dynamism of the disease, potentially providing a method very useful for patient management during treatment. The detection of ctDNA in advanced lung cancer stages has already been demonstrated in many studies, and concordance between ctDNA and tissue increases with increasing TNM staging (i.e., stage I, 57.9%; stage II, 66.7%; stage IIIA, 90%). Furthermore, the ctDNA proved to be more sensitive than traditional tumor markers, including CA19.9 and CEA15.5, in various studies. The analysis of the ctDNA can provide information about the intra-tumoral heterogeneity, which contributes, through clonal selection, to the appearance of resistance mechanism to treatment [53]. It has been shown that some targetable mutations, such as *EGFR*, *MET*, and *BRAF* are mostly mutually exclusive at the clonal level and appear early in tumorigenesis, thus explaining the important responses seen in different disease sites when these alterations are targeted with specific drugs [54,55]. During therapy, more than 75% of tumors develop a clonal selection of driver alterations, with the appearance of new mutations in genes such as *PIK3CA*, *NF1*, *KRAS*, *TP53*, and *NOTCH* [56]. Furthermore, some mutations may appear during therapy at specific sites, confirming the limitations of a single site tissue biopsy [56]. ctDNA analysis plays a very important role in monitoring response to treatment in NSCLC patients with oncogene-addicted tumors. Several studies have shown that ctDNA in mutated *EGFR* patients is useful for:

- detect both the activating *EGFR* mutations and the presence of any other mutations at baseline, allowing thus quantitative monitoring of dynamic alterations during treatment with TKIs (gefitinib, erlotinib, afatinib, osimertinib) [57—59].
- identify the appearance of new resistance mutations simultaneously with progression but earlier with respect to radiological evidence [60,61].

Furthermore, in patients harboring ALK translocation, ctDNA monitoring during the treatment with TKI (crizotinib, alectinib, brigatinib, ceritinib, lorlatinib) can be a useful tool for:

- identify the appearance of new resistance mutations in the *ALK* gene or in other genes [62,63];
- monitor the evolution of secondary *ALK* mutations or other gene alterations during treatment [62,63].

Among the major limitations of ctDNA monitoring, it remains the impossibility of identifying resistance mechanisms associated with histotype switch, and a guideline that identifies the most suitable technology for the analysis [64].

In conclusion, the information supporting the use of LB for monitoring targeted therapy in NSCLC (mutated *EGFR* or translocated *ALK*) is solid and reproducible. Considering the possibility of using subsequent treatment lines depending on well-differentiated molecular characteristics, ctDNA monitoring represents a tool compatible with minimally invasive as well as with a sensitivity and specificity suitable for the purpose.

4.3.2 Monitoring of treatment response in colorectal cancer

LB has demonstrated high feasibility in CRC for the evaluation of the molecular dynamism by catching temporal biological heterogeneity of the tumor, even overcoming the limits of spatial heterogeneity of tissue biopsy. The search for *RAS* mutations in metastatic colorectal disease, in fact, aims to monitor the acquisition of resistance during treatment with anti-EGFR as well as the refinement of patients' selection for therapeutic rechallenge. In both cases, we rely on strong translational evidence that has shown that resistance to anti-EGFR is associated with the increase of mutated *RAS* alleles and the appearance of mutation in the *EGFR* extracellular domain [65,66]. Furthermore, emerging data are available on the use of LB in the determination of *RAS*-independent resistance mechanisms, such as example the amplification of *MET* or *ERBB2* and *BRAF* and *MEK* mutations, opening the scenario for new therapeutic perspectives [65]. However, the data supporting the use of LB for monitoring anti-EGFR therapy are limited to small patient cohorts and are not conclusive for immediate use in clinical practice [67], lacking thus evidence of any therapeutic impact. The data regarding molecular selection for rechallenge are strongly suggestive of the possible and useful use of such a therapeutic strategy, even if it still requires validation on larger populations [67–70].

4.4 Key points

1. In current clinical practice, LB is mainly used to analyze EGFR gene mutational status in advanced NSCLC patients;
2. Point mutation in BRAF gene, rearrangements of ALK, ROS1, and NTRK genes as well as MET exon 14 skipping, MET and HER2 amplification, and the p.G12C mutation in KRAS genes are all involved in positively or negatively defining treatment response;
3. ctDNA analysis represents a valuable alternative whenever tissue samples are not suitable for detecting EGFR mutational status;
4. All *EGFR*-mutant patients that received first- or second-generation TKIs (gefitinib, erlotinib o afatinib) as first-line therapy, experiencing disease progression, must be tested for the mutation p.T790M arising in *EGFR* exon 20;
5. The ability to use ctDNA as a marker of MRD has become an emerging area of clinical research for localized CRC;
6. *RAS* liquid testing as a potential substitute for tumor tissue analysis in metastatic CRC;
7. LB analysis of frequently mutated genes in breast cancer (i.e., ESR1, PIK3CA) has been matched with the evaluation of tumor burden, the identification and monitoring of residual disease in patients undergoing curative surgery, and as a significant prognostic factor;
8. LB could represent a useful tool in melanoma patients, both for the identification of *BRAF* and *NRAS* mutations to set up the treatment (if the tissue is not available), and for the quantitative monitoring of the ctDNA during treatment.

Acknowledgments

M. La Mantia and V. Gristina contributed to the current work under the Doctoral Program in Experimental Oncology and Surgery, University of Palermo.

References

[1] Rolfo C, Mack PC, Scagliotti GV, et al. Liquid biopsy for advanced non-small cell lung cancer (NSCLC): a statement paper from the IASLC. J Thorac Oncol 2018;13 (9):1248–68.
[2] Mok TS, Wu Y-L, Ahn M-J. Osimertinib or platinum-pemetrexed in EGFR T790M-positive lung cancer. N Engl J Med 2017;376(7):629–40.

[3] Passiglia F, Rizzo S, Rolfo C, et al. Metastatic site location influences the diagnostic accuracy of ctDNA EGFR- mutation testing in NSCLC patients: a pooled analysis. Curr Cancer Drug Targets 2018;18(7):697–705.

[4] Passiglia F, Pilotto S, Facchinetti F, et al. Treatment of advanced non-small-cell lung cancer: the 2019 AIOM (Italian Association of Medical Oncology) clinical practice guidelines. Crit Rev Oncol Hematol 2020;146:102858.

[5] Guibert N, Pradines A, Favre G, et al. Current and future applications of liquid biopsy in non-small cell lung cancer from early to advanced stages. Eur Respir Rev 2020;29(155):190052.

[6] Vogelstein B, Kinzler KW. Digital PCR. Proc Natl Acad Sci U S A 1999;96 (16):9236–41.

[7] Zhang BO, Xu CW, Shao Y, et al. Comparison of droplet digital PCR and conventional quantitative PCR for measuring EGFR gene mutation. Exp Ther Med 2015;9 (4):1383–8.

[8] Gérin M, Siemiatycki J, Nadon L, et al. Cancer risks due to occupational exposure to formaldehyde: results of a multi-site case-control study in Montreal. Int J Cancer 1989;44(1):53–8.

[9] Esposito Abate R, Pasquale R, Fenizia F, et al. The role of circulating free DNA in the management of NSCLC. Expert Rev Anticancer Ther 2019;19(1):19–28.

[10] Gristina V, Malapelle U, Galvano A, et al. The significance of epidermal growth factor receptor uncommon mutations in non-small cell lung cancer: a systematic review and critical appraisal. Cancer Treat Rev 2020;85:101994.

[11] Malapelle U, Pisapia P, Rocco D, et al. Next-generation sequencing techniques in liquid biopsy: focus on non-small cell lung cancer patients. Transl Lung Cancer Res 2016;5(5):505–10.

[12] Remon J, Lacroix L, Jovelet C, et al. Real-world utility of an amplicon-based next-generation sequencing liquid biopsy for broad molecular profiling in patients with advanced non-small-cell lung cancer. JCO Precis Oncol 2019;3 PO.18.00211.

[13] Sabari JK, Offin M, Stephens D, et al. A prospective study of circulating tumor DNA to guide matched targeted therapy in lung cancers. J Natl Cancer Inst 2019;111(6):575–83.

[14] Kato S, Schwaederlé MC, Fanta PT, et al. Genomic assessment of blood-derived circulating tumor DNA in patients with colorectal cancers: correlation with tissue sequencing, therapeutic response, and survival. JCO Precis Oncol 2019;3 PO.18.00158.

[15] Kilgour E, Rothwell DG, Brady G, Dive C. Liquid biopsy-based biomarkers of treatment response and resistance. Cancer Cell 2020;37(4):485–95.

[16] Parikh AR, Leshchiner I, Elagina L, et al. Liquid vs tissue biopsy for detecting acquired resistance and tumor heterogeneity in gastrointestinal cancers. Nat Med 2019;25(9):1415–21.

[17] Lin KK, Harrell MI, Oza AM, et al. BRCA reversion mutations in circulating tumor DNA predict primary and acquired resistance to the PARP inhibitor rucaparib in high-grade ovarian carcinoma. Cancer Discov 2019;9(2):210–19.

[18] Tie J, Wang Y, Tomasetti C, et al. Circulating tumor DNA analysis detects minimal residual disease and predicts recurrence in patients with stage II colon cancer. Sci Transl Med 2016;8(346) 346ra92.

[19] Reinert T, Henriksen TV, Christensen E, et al. Analysis of plasma cell-free DNA by ultradeep sequencing in patients with stages I to III colorectal cancer. JAMA Oncol 2019;5(8):1124–31.

[20] Tarazona N, Gimeno-Valiente F, Gambardella V, et al. Targeted next-generation sequencing of circulating-tumor DNA for tracking minimal residual disease in localized colon cancer. Ann Oncol 2019;30(11):1804–12.

[21] Normanno N, Cervantes A, Ciardiello F, De Luca A, Pinto C. The liquid biopsy in the management of colorectal cancer patients: current applications and future scenarios. Cancer Treat Rev 2018;70:1–8 Nov.

[22] Hale AR. Stress, werkdruk en fouten [Stress, work pressures and mistakes]. TVZ. 1988;42(21):673–5 Dutch. PMID: [25].

[23] O'Leary B, Hrebien S, Morden JP, et al. Early circulating tumor DNA dynamics and clonal selection with palbociclib and fulvestrant for breast cancer. Nat Commun 2018;9(1):896.

[24] Alimirzaie S, Bagherzadeh M, Akbari MR. Liquid biopsy in breast cancer: a comprehensive review. Clin Genet 2019;95(6):643–60.

[25] Majure M, Logan AC. What the blood knows: interrogating circulating tumor DNA to predict progression of minimal residual disease in early breast cancer. Ann Transl Med 2016;4(24):543.

[26] Wang P, Bahreini A, Gyanchandani R, et al. Sensitive detection of mono- and polyclonal ESR1 mutations in primary tumors, metastatic lesions, and cell-free DNA of breast cancer patients. Clin Cancer Res 2016;22(5):1130–7.

[27] Beije N, Sieuwerts AM, Kraan J, et al. Estrogen receptor mutations and splice variants determined in liquid biopsies from metastatic breast cancer patients. Mol Oncol 2018;12(1):48–57.

[28] Zhou Y, Xu Y, Gong Y, et al. Clinical factors associated with circulating tumor DNA (ctDNA) in primary breast cancer. Mol Oncol 2019;13(5):1033–46.

[29] Lee JH, Jeong H, Choi JW, Oh HE, Kim YS. Liquid biopsy prediction of axillary lymph node metastasis, cancer recurrence, and patient survival in breast cancer: a meta-analysis. Med (Baltim) 2018;97(42):e12862.

[30] André F, Ciruelos E, Rubovszky G, et al. SOLAR-1 study group. Alpelisib for PIK3CA-mutated, hormone receptor-positive advanced breast cancer. N Engl J Med 2019;380(20):1929–40.

[31] Ma CX, Crowder RJ, Ellis MJ. Importance of PI3-kinase pathway in response/resistance to aromatase inhibitors. Steroids. 2011;76(8):750–2.

[32] Araki K, Miyoshi Y. Mechanism of resistance to endocrine therapy in breast cancer: the important role of PI3K/Akt/mTOR in estrogen receptor-positive, HER2-negative breast cancer. Breast Cancer 2018;25(4):392–401.

[33] Boyer M, Cayrefourcq L, Dereure O, et al. Clinical relevance of liquid biopsy in melanoma and merkel cell carcinoma. Cancers (Basel) 2020;12(4):960 [36]; PMCID: PMC7226137.

[34] Syeda MM, Wiggins JM, Corless B, et al. Validation of circulating tumor DNA assays for detection of metastatic melanoma. Methods Mol Biol 2020;2055:155–80.

[35] Diefenbach RJ, Lee JH, Rizos H. Monitoring melanoma using circulating free DNA. Am J Clin Dermatol 2019;20(1):1–12.

[36] Pinzani P, Salvianti F, Zaccara S, et al. Circulating cell-free DNA in plasma of melanoma patients: qualitative and quantitative considerations. Clin Chim Acta 2011;412 (23–24):2141.

[37] Herbreteau G, Charpentier S, Vallée A, Denis MG. Use of circulating tumoral DNA to guide treatment for metastatic melanoma. Pharmacogenomics. 2019;20 (18):1259–70.

[38] Santiago-Walker A, Gagnon R, Mazumdar J, et al. Correlation of BRAF mutation status in circulating-free DNA and tumor and association with clinical outcome across four BRAFi and MEKi clinical trials. Clin Cancer Res 2016;22 (3):567–74.

[39] Rowe SP, Luber B, Makell M, et al. From validity to clinical utility: the influence of circulating tumor DNA on melanoma patient management in a real-world setting. Mol Oncol 2018;12(10):1661–72.

[40] Lee RJ, Gremel G, Marshall A, et al. Circulating tumor DNA predicts survival in patients with resected high-risk stage II/III melanoma. Ann Oncol 2018;29 (2):490−6.

[41] Lee JH, Saw RP, Thompson JF, et al. Pre- operative ctDNA predicts survival in high-risk stage III cutaneous melanoma patients. Ann Oncol 2019;30(5):815−22.

[42] Tan L, Sandhu S, Lee RJ, Li J, et al. Prediction and monitoring of relapse in stage III melanoma using circulating tumor DNA. Ann Oncol 2019;30(5):804−14.

[43] Gray ES, Rizos H, Reid AL, et al. Circulating tumor DNA to monitor treatment response and detect acquired resistance in patients with metastatic melanoma. Oncotarget 2015;6(39):42008−18.

[44] Cabel L, Riva F, Servois V, et al. Circulating tumor DNA changes for early monitoring of anti-PD1 immunotherapy: a proof-of-concept study. Ann Oncol 2017;28 (8):1996−2001.

[45] Schreuer M, Meersseman G, Van Den Herrewegen S, et al. Quantitative assessment of BRAF V600 mutant circulating cell-free tumor DNA as a tool for therapeutic monitoring in metastatic melanoma patients treated with BRAF/MEK inhibitors. J Transl Med 2016;14:95.

[46] Gonzalez-Cao M, Mayo de Las Casas C, Jordana Ariza N, Manzano JL, et al. Early evolution of BRAFV600 status in the blood of melanoma patients correlates with clinical outcome and identifies patient's refractory to therapy Spanish Melanoma Group Melanoma Res 2018;28(3):195−203.

[47] Lin SY, Huang SK, Huynh KT, et al. Multiplex gene profiling of cell-free DNA in patients with metastatic melanoma for monitoring disease. JCO Precis Oncol 2018;2 PO.17.00225.

[48] Del Re M, Marconcini R, Pasquini G, et al. PD-L1 mRNA expression in plasma-derived exosomes is associated with response to anti-PD-1 antibodies in melanoma and NSCLC. Br J Cancer 2018;118(6):820−4.

[49] Knuever J, Weiss J, Persa OD, et al. The use of circulating cell-free tumor DNA in routine diagnostics of metastatic melanoma patients. Sci Rep 2020;10(1):4940.

[50] Knol AC, Vallée A, Herbreteau G, et al. Clinical significance of BRAF mutation status in circulating tumor DNA of metastatic melanoma patients at baseline. Exp Dermatol 2016;25(10):783−8.

[51] Greaves M. Evolutionary determinants of cancer. Cancer Discov 2015;5(8):806−20.

[52] Mok TS, Wu YL, Thongprasert S, et al. Gefitinib or carboplatin-paclitaxel in pulmonary adenocarcinoma. N Engl J Med 2009;361(10):947−57.

[53] Cao Y, Xiao G, Qiu X, Ye S, Lin T. Efficacy and safety of crizotinib among Chinese EML4-ALK-positive, advanced-stage non-small cell lung cancer patients. PLoS One 2014;9(12):e114008.

[54] Jamal-Hanjani M, Wilson GA, McGranahan N, Birkbak NJ, et al. TRACERx consortium. tracking the evolution of non-small-cell lung cancer. N Engl J Med 2017;376(22):2109−21.

[55] Del Re M, Rofi E, Cappelli C, et al. The increase in activating EGFR mutation in plasma is an early biomarker to monitor response to osimertinib: a case report. BMC Cancer 2019;19(1):410.

[56] Buttitta F, Felicioni L, Lorito AD, et al. Early prediction of resistance to tyrosine kinase inhibitors by plasma monitoring of EGFR mutations in NSCLC: a new algorithm for patient selection and personalized treatment. Oncotarget. 2020;11 (11):982−91.

[57] Marchetti A, Palma JF, Felicioni L, et al. Early prediction of response to tyrosine kinase inhibitors by quantification of EGFR mutations in plasma of NSCLC patients. J Thorac Oncol 2015;10(10):1437−43.

[58] Gray JE, Okamoto I, Sriuranpong V, et al. Tissue and plasma EGFR mutation analysis in the FLAURA trial:osimertinib vs comparator EGFR Tyrosine kinase inhibitor as first-line treatment in patients with EGFR-mutated advanced non-small cell lung cancer. Clin Cancer Res 2019;25(22):6644−52.

[59] Papadimitrakopoulou VA, Han JY, Ahn MJ, et al. Epidermal growth factor receptor mutation analysis in tissue and plasma from the AURA3 trial: osimertinib vs platinum-pemetrexed for T790M mutation-positive advanced non- small cell lung cancer. Cancer. 2020;126(2):373−80.

[60] Bordi P, Tiseo M, Rofi E, et al. Detection of ALK and KRAS Mutations in circulating tumor DNA of patients with advanced ALK-Positive NSCLC With disease progression during crizotinib treatment. Clin Lung Cancer 2017;18(6):692−7.

[61] Dagogo-Jack I, Brannon AR, Ferris LA, Campbell CD, et al. Tracking the evolution of resistance to ALK tyrosine kinase inhibitors through longitudinal analysis of circulating tumor DNA. JCO Precis Oncol 2018;2018 PO.17.00160.

[62] Del Re M, Crucitta S, Gianfilippo G, Passaro A, et al. Understanding the mechanisms of resistance in EGFR-positive NSCLC: from tissue to liquid biopsy to guide treatment strategy. Int J Mol Sci 2019;20(16):3951.

[63] Siravegna G, Mussolin B, Buscarino M, et al. Clonal evolution and resistance to EGFR blockade in the blood of colorectal cancer patients. Nat Med 2015;21(7):827.

[64] Parseghian CM, Loree JM, Morris VK, et al. Anti-EGFR-resistant clones decay exponentially after progression: implications for anti-EGFR re-challenge. Ann Oncol 2019;30(2):243−9.

[65] Siena S, Sartore-Bianchi A, Garcia-Carbonero R, et al. Dynamic molecular analysis and clinical correlates of tumor evolution within a phase II trial of panitumumab-based therapy in metastatic colorectal cancer. Ann Oncol 2018;29(1):119−26.

[66] Cremolini C, Rossini D, Dell'Aquila E, et al. Rechallenge for patients with RAS and BRAF wild-type metastatic colorectal cancer with acquired resistance to first-line cetuximab and irinotecan: a phase 2 single-arm clinical trial. JAMA Oncol 2019;5(3):343−50.

[67] Mauri G, Pizzutilo EG, Amatu A, et al. Retreatment with anti-EGFR monoclonal antibodies in metastatic colorectal cancer: systematic review of different strategies. Cancer Treat Rev 2019;73:41−53.

[68] Russo A, Incorvaia L, Del Re M, Malapelle U, et al. The molecular profiling of solid tumors by liquid biopsy: a position paper of the AIOM-SIAPEC-IAP-SIBioC-SIC-SIF Italian Scientific Societies. ESMO Open 2021;6(3):100164.

[69] Russo A, Incorvaia L, Capoluongo E, Tagliaferri P, Galvano A, Del Re M, et al. The challenge of the Molecular Tumor Board empowerment in clinical oncology practice: a position paper on behalf of the AIOM- SIAPEC/IAP-SIBioC-SIC-SIF-SIGU-SIRM Italian Scientific Societies. Crit Rev Oncol Hematol 2022;169:103567. Available from: https://doi.org/10.1016/j.critrevonc.2021.103567 Epub 2021 Dec 8. PMID: 34896250.

[70] Passiglia F, Galvano A, Gristina V, et al. Is there any place for PD-1/CTLA-4 inhibitors combination in the first-line treatment of advanced NSCLC?-a trial-level meta-analysis in PD-L1 selected subgroups. Transl Lung Cancer Res 2021;10(7):3106−19. Available from: https://doi.org/10.21037/tlcr-21-52.

CHAPTER 5

Liquid biopsy and immunotherapy: is all that glitter gold?

L. Incorvaia[1], A. Perez[1], C. Brando[1], V. Gristina[1], M. La Mantia[1], M. Castiglia[1], D. Fanale[1], A. Galvano[1], G. Badalamenti[1], A. Russo[1] and V. Bazan[2]

[1]Department of Surgical, Oncological, and Oral Sciences, University of Palermo, Palermo, Italy
[2]Department of Biomedicine, Neuroscience and Advanced Diagnostics (Bi.N.D.), University of Palermo, Palermo, Italy

Learning objectives

By the end of the chapter, the reader will

1. Have learned the basic concept of liquid biopsy (LBs) and its application in oncology.
2. Have reached knowledge about Circulating tumor cells (CTCs), cell-free DNA (cfDNA), Exosomes, T-lymphocytes [the T-cell receptor (TCR)], Cytokines, soluble forms of Immune-Checkpoints and other circulating proteins.
3. Have known clinical trials investigating LBs in immunotherapy in different tumor types.

5.1 Background: the need for predictive biomarkers for patient selection

Immuno-oncology (IO), particularly immune checkpoint inhibitors (ICIs), is rapidly evolving in the treatment of several human cancers by enhancing and increasing the ability of the immune system to generate an antitumor response against cancer cells through natural mechanisms that are evaded or damaged during disease progression [1,2]. The introduction of ICIs has become a promising frontier for the treatment of several solid tumors, such as the non–small-cell lung cancer (NSCLC), melanoma [3–7], renal cell carcinoma (RCC) [8], but also small–cell lung cancer (SCLC) [9] and breast [10], cervical [11], bladder [12–14], colorectal [15], gastric [16] and liver [17] tumors.

Liquid Biopsy
DOI: https://doi.org/10.1016/B978-0-12-822703-9.00005-3

© 2023 Elsevier Inc.
All rights reserved.

However, despite many patients showing a durable response to ICIs, the treatment efficacy is variable and poorly predictable in daily clinical practice. In addition, such IO drugs, requiring a high dose because of their short half-life, cause severe autoimmune side effects [2]. In this complex scenario, the variability of clinical response to immunotherapy requires the discovery of predictive biomarkers for patient selection.

Since ICIs inhibit the PD-1/PD-L1 (programmed death receptors 1/programmed death receptor-ligand 1) interactions, the assessment of PD-L1 expression using immunohistochemistry (IHC) staining in formalin-fixed paraffin-embedded (FFPE) tissue samples was investigated as a predictive biomarker. The results were heterogeneous in the different types of tumors. IHC technical methods and biological heterogeneity within the tumor sample are the main limitations to using PD-L1 expression as predictive biomarkers to identify patients who will likely benefit from immunotherapy. Thus, there is intensive work to develop more dynamic biomarkers for therapeutic decisions in the clinic.

Several components, using LB, are under-investigation as predictive biomarkers of immunotherapy response or resistance, such as CTCs, cfDNA, and Exosomes, but also T-lymphocytes (the TCR), Cytokines, soluble form of Immune-Checkpoints and other circulating proteins (Fig. 5.1).

5.1.1 Circulating tumor cells, cell-free DNA, and exosomes

As previously mentioned, the evaluation of PD-L1 expression by IHC represents the only predictive biomarker used as a companion diagnostic test for first-line immunotherapy with anti-PD-L1 or anti-PD1 drugs. However, there are several technical and biological issues and PD-L1 assessment could be sometimes challenging [18]. One of the main drawbacks is PD-L1 heterogenous expression within the same lesion and among different tumor sites [19]. This implies that PD-L1 status could be misinterpreted or underestimated especially when the analysis is conducted in small biopsies instead of surgical resection. Test misinterpretation could, ultimately, lead to patient withdrawal from immunotherapeutic treatment options, thus profoundly impairing their treatment chances. Consequently, there is an urgent need to identify new tools for tumor characterization that overcomes the aforementioned issues and that provide also the opportunity for a strict follow-up of patients during treatment. In particular, to obtain patients' serial monitoring it is very

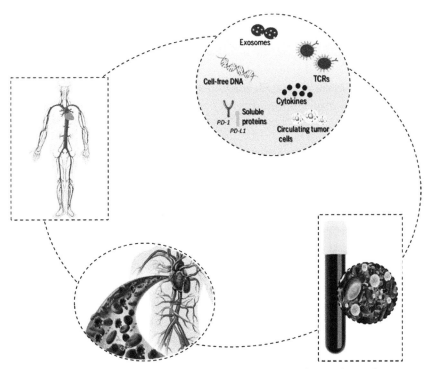

Figure 5.1 Components under-investigation as predictive biomarkers of immunotherapy response or resistance, using liquid biopsy.

important to use minimally invasive and simple procedures that can be performed in an outpatient facility. LB encompasses all the demanded needs: it is minimally invasive; it is easily repeatable over time; it allows to bypass of issues related to tumor heterogeneity; it is time- and cost-effective.

LB utilizes different body fluids, mainly blood, to obtain tumor-derived material that can be used for biological testing [20]. By using LB we can collect CTCs, circulating nucleic acids such as circulating cell-free DNA (ccfDNA), or more specifically circulating tumor DNA (ctDNA) but also circulating vesicles like exosomes [21]. LB utility in a clinical setting has been, so far, mainly investigated in the context of targeted therapies, especially in thoracic oncology. Indeed, in Advanced NSCLC the analysis of *EGFR* mutational status can be conducted on LB; clinicians can therefore use LB as a decision-making strategy for treatment choice [22,23]. Given these encouraging results and considering the unmet needs for new surrogate and predictive biomarkers for immunotherapeutic regimens, there is an increasing interest in bringing LB into the domain of immunotherapy. In this context, LB can be used as a predictive biomarker at baseline or during treatment to monitor the onset of resistance to immunotherapy. In Table 5.1 there is a list of all clinical trials retrieved from the website https://clinicaltrials.gov that aim at the investigation of LB in the context of immunotherapy.

5.1.2 Circulating tumor cells

CTCs are cancer cells that detach from a solid tumor lesion and enter into circulation. During their journey in the bloodstream, they can be easily captured with a simple blood withdrawal. Given their origin, CTCs normally share several biological and cytological characteristics with the lesion from which they are generated and are therefore perfect candidates for biological testing.

Compared to ctDNA, CTCs are suitable to perform analysis not only at a molecular level but also at protein and cellular levels. On the other hand, CTCs are challenging to isolate and may not promptly reflect tumor evolution as well as ctDNA. However, since the only validated PD-L1 test is based on the analysis of its expression at a protein level, CTCs may represent a useful dynamic source to obtain this kind of information not only at baseline but also during treatment. Moreover, CTCs may provide a global portrait of PD-L1 status on different tumor sites, especially when considering that immune checkpoint protein expression is highly heterogenous and can be influenced by multiple microenvironmental, inflammatory and therapeutic factors [24].

Most of the existing approaches for CTCs analysis are based on a two-step process that starts with cell enrichment followed by detection. CTCs capturing can be obtained by exploiting CTCs' physical properties (charge, size, elasticity, and density) [25−27] as well as various biological features (cellular functions [28] and tumor-specific surface proteins [29,30]).

The first step to prove that CTCs can be used for PD-L1 status evaluation was to demonstrate that PD-L1 is expressed on CTCs. Consequently, in 2015 the group Mazel et al. demonstrated for the first time CTCs PD-L1 expression in patients with hormone receptor-positive, HER2-negative breast cancer [31]. In this paper, CTCs isolation was performed using the EpCAM-based CellSearch system, which remains still the only Food and Drug Administration (FDA)-approved device for

Table 5.1 List of clinical trials retrieved from https://clinicaltrials.gov investigating LB in immunotherapy in different tumor types at different stages.

Study title	Cancer type(s)	Biomarker	Treatment	Study ai(s)	Study ID
Dynamic Biomarkers of Immune Microenvironment for Stage III Locally Advanced Unresectable Non-small-cell Lung Cancer	NSCLC	PBMC and ctDNA	Chemoradio-therapy	– Biomarkers associated with immune efficacy – The optimal time for immunotherapy	NCT04749407
Prospective Study for the Prognostic and Predictive Role of Circulating Tumor Cells in Patients With Oropharyngeal Advanced Squamous Cell Carcinoma: CTCO (Circulating Tumor Cells in the Oropharynx)	Advanced squamous cell carcinoma of the oropharynx	CTC	Curative treatment	– CTCs characterization	NCT04696744
Establishment of a Comprehensive Evaluation System for the Efficacy of Immunologic Checkpoint Inhibitors in the Treatment of Advanced Non-small Cell Lung Cancer	Advanced NSCLC	CTC	Immune checkpoint inhibitors	– CTCs characterization – T cell marker analysis	NCT04629027

(*Continued*)

Table 5.1 (Continued)

Study title	Cancer type(s)	Biomarker	Treatment	Study ai(s)	Study ID
A Clinical Performance Study to Validate the Use of Novel Molecular Diagnostic Assays for the Detection of Cancer Biomarkers in Peripheral Blood and Primary Tumor Tissue of Patients With Recurrent/Metastatic HNSCC, Metastatic NSCLC, or Metastatic Melanoma	HNSCC, NSCLC, melanoma	CTC	Nivolumab, Pembrolizumab	– Clinical performance of PD-L1 kit in CTCs of peripheral blood and tumor tissue samples	NCT04490564
Predictive Biomarkers For Response To Nivolumab In Head and Neck Squamous Cell Carcinoma	HNSCC	CTC	Nivolumab	– Change in the percentage of immune cells post-treatment compared to baseline biopsies – PD-L1 expression in CTCs in association with best overall response and survival	NCT03652142
WHENII - Early Response Evaluation With FDG-PET/CT and Liquid Biopsy in Patients With NSCLC	NSCLC	CTC, ctDNA	Immunotherapy or chemotherapy	– Observation of changes in cancer cells metabolism during 1st treatment with chemotherapy or immunotherapy – Observation of changes in circulating tumor DNA during 1st treatment	NCT03481101

Identification and Evaluation of the Potential Biomarkers on Circulating Tumor Cells and Tumor Related Rare Cells in Cancer Patients Undergoing Immunotherapy	Cancer patients on immune checkpoint inhibitors	CTCs	Immune checkpoint inhibitors	— Isolation and analysis of circulating tumor cells before and after immunotherapy	NCT03434912
I-CURE-1: A Phase II, Single-Arm Study of Pembrolizumab Combined With Carboplatin in Patients With Circulating Tumor Cells (CTCs) Positive Her-2 Negative Metastatic Breast Cancer (MBC)	HER2 negative MBC	CTCs, ctDNA	Pembrolizumab + carboplatin	— Evaluate PFS in CTC +, HER2- MBC patients treated with a combination of pembrolizumab + carboplatin and previously treated with anthracyclines and taxanes — Measure PDL-1 in CTCs (CellSearch) and immune cells such as cancer-associated macrophage-like cells (CAMLs) (CellSieve) and correlate with therapeutic benefit — Measure ctDNA and T-cell receptor sequencing analysis and correlate them with CTC enumeration and therapeutic benefit	NCT03213041

(Continued)

Table 5.1 (Continued)

Study title	Cancer type(s)	Biomarker	Treatment	Study ai(s)	Study ID
Analysis of Circulating Tumor Markers in the Blood (ALCINA)	Any tumor type	Multiple	Multiple treatments, including immunotherapy	— Feasibility of the analysis of different blood-borne tumor biomarkers — Correlation with biological and clinical data	NCT02866149
PDL-1 Expression on Circulating Tumor Cells in Non-small Cell Lung Cancer (IMMUNO-PREDICT)	NSCLC	CTCs, MDSC	Immunotherapy	— Feasibility of analysis of PDL-1 expression on CTC as assessed by immunocytochemistry analysis — Percentage of CTC expressing PD-L1 after four cycles of immunotherapy as assessed by immunocytochemistry analysis — Evolution of MDSCs count in response to treatment as assessed by MDSCs analysis	NCT02827344
Defining the Relevant Immune Checkpoints Expressed on Metastatic Prostate Cancer Circulating Tumor Cells	mCRPC	CTCs	Sipuleucel-T, abiraterone acetate, enzalutamide	— Change in expression of four immune checkpoint biomarkers (PD-L1, PD-L2, B7-H3, and CTLA-4) on CTCs	NCT02456571

Title	Cancer type	Biomarker	Treatment	Objectives	NCT number
The Clinical Relevance of Immune Cells and Circulating Tumor Cells in Patients With Head and Neck Cancer	HNSCC	CTCs	Palliative chemotherapy	— The correlation among CTC numbers, PD-1 expressions, and the prognosis of patients	NCT02420600
Study on the Prediction of Immunotherapeutic Effect of Advanced Non-small Cell Lung Cancer by Detection of Plasma Exosomes	NSCLC	Exosomes, circulating miRNAs	anti-PD-1	— PD-L1 expression profiles of exosomes — miRNA expression profiles of exosomes	NCT04427475
Exosomes and Resistance to Immunotherapy in Aggressive Non-Hodgkin B-cell Lymphomas (B-NHL)	DLBCL	Exosomes	Rituximab (anti-CD20)	— Quantification of CD20 and PDL-1 in exosomes — Analyze the prognostic value of exosomal markers on therapeutic response and patient outcome	NCT03985696
Clinical Research for the Consistency Analysis of PD-L1 in Cancer Tissue and Plasma Exosome (RadImm01)	NSCLC	Exosomes	No treatment	— Concordance between PD-L1 protein expression in cancer tissue and PD-L1 mRNA expression in exosomes — Concordance between PD-L1 protein expression and PD-L1 mRNA in tumor tissue	NCT02890849

(Continued)

Table 5.1 (Continued)

Study title	Cancer type(s)	Biomarker	Treatment	Study aim(s)	Study ID
Clinical Research for the Consistency Analysis of PD-L1 in Lung Cancer Tissue and Plasma Exosome Before and After Radiotherapy (RadImm02)	NSCLC	Exosomes	Radiotherapy	– Concordance between PD-L1 protein expression in cancer tissue and PD-L1 mRNA expression in exosomes before and after radiotherapy – to identify the best radiotherapy division and timing which can make PD-L1 express more	NCT02869685
A Trial Using ctDNA Blood Tests to Detect Cancer Cells After Standard Treatment to Trigger Additional Treatment in Early-Stage Triple-Negative Breast Cancer Patients (c-TRAK-TN)	TNBC	ctDNA	Pembrolizumab	– To assess whether ctDNA screening can be used to detect MRD following a patient's standard primary treatment for TNBC – To evaluate ctDNA clearance after 6 months after commencing pembrolizumab	NCT03145961
An Observational Study of Circulating Tumor DNA Genetic Alterations in Non-small Cell Lung Cancer Patients Treated With Pembrolizumab	NSCLC	ctDNA	Pembrolizumab	– Radiologic response to immune checkpoint blockade by clonal dynamics of serial ctDNA in 1st line NSCLC patients receiving pembrolizumab monotherapy	NCT04791215

BESPOKE Study of ctDNA Guided Immunotherapy	Melanoma, NSCLC, CRC	ctDNA	Immune checkpoint inhibitors	– Examine the impact of SIGNATERA on treatment decisions on tumor assessment timepoints after initiation of immunotherapy	NCT04761783
A Prospective Study on NGS-based Comprehensive Genomic ctDNA Panel in NSCLC Treated With Immunotherapy	NSCLC	ctDNA, bTMB	Immunotherapy	– To investigate the combined predictive biomarkers (including bTMB and HLA) related to the immunotherapy effects and the biomarker (TCR) associated with adverse reactions during immunotherapy and hold a predictive role	NCT04636047
Study of Circulating Tumor DNA (ctDNA) Kinetics in Immuno-oncology: Intense Dynamic Monitoring of ctDNA in Advanced/ Metastatic Head and Neck Squamous Cell Carcinoma (HNSCC) Patients Treated	HNSCC	ctDNA	Nivolumab or pembrolizumab	– To evaluate the change in the kinetics of ctDNA in advanced/metastatic HNSCC patients treated with immune checkpoint inhibitors – To correlate the changes in ctDNA levels with PFS and OS	NCT04606940

(Continued)

Table 5.1 (Continued)

Study title	Cancer type(s)	Biomarker	Treatment	Study aim(s)	Study ID
With Immune Checkpoint Inhibitors				– To evaluate the optimal time-point to analyze ctDNA as a predictive marker of response to immune checkpoint inhibitors	
Treatment Of Metastatic Bladder Cancer at the Time Of Biochemical reLApse Following Radical Cystectomy (TOMBOLA)	Bladder	ctDNA	Atezolizumab	– Complete response (CR) after treatment with investigational agent initiated by ctDNA positive status after radical cystectomy	NCT04138628
Prediction of the Efficacy of ctDNA in Immunotherapy for Advanced Gastric Cancer	Gastric cancer	ctDNA	Immune checkpoint inhibitors	– To explore the possibility of the clinical utility of serum ctDNA as a clinical index to predict the efficacy in immunotherapy for advanced gastric cancer	NCT04053725
Early Assessment of Response to Treatment of Metastatic LUng Tumors Based on CIrculating Tumor DNA (ELUCID)	NSCLC	ctDNA	Targeted therapies, immunotherapy, chemotherapy	– To determine whether early evolution (between baseline and week 3) of ctDNA concentration predicts the radiological response to 1st-line treatment of advanced or metastatic NSCLC patients	NCT03926260

Study	Cancer	Biomarker	Treatment	Objectives	NCT
CAcTUS - Circulating Tumor DNA Guided Switch (CAcTUS)	Melanoma	ctDNA	Targeted therapies, immunotherapy	– To determine whether switching from targeted therapy to immunotherapy based on a decrease in levels of ctDNA improve the outcome in melanoma patients	NCT03808441
Ipilimumab and Nivolumab in Recurrent Extensive Stage Small Cell Lung Cancer After Receiving Platinum–based Chemotherapy	SCLC	ctDNA, Teff/Treg ratio	Ipilimumab and Nivolumab	– To assess whether the change of Teff/Treg ratio between pre- and on-treatment biopsies, will predict clinical response in patients with SCLC treated with nivolumab and ipilimumab – To assess if the change in ctDNA variant allele frequencies from the baseline sample to the 4-week sample predicts response to therapy and correlates to the objective response rate	NCT03670056
Anti-PD-1 Alone or Combined With Autologous Cell Therapy in Advanced NSCLC	NSCLC	ctDNA	Pembrolizumab	– To investigate the relationship of ctDNA with clinical outcomes	NCT03360630

(Continued)

Table 5.1 (Continued)

Study title	Cancer type(s)	Biomarker	Treatment	Study aim(s)	Study ID
Selection Pressure and Evolution Induced by Immune Checkpoint Inhibitors and Other Immunologic Therapies (SPECIAL)	Head and neck, melanoma	ctDNA	Immune checkpoint inhibitors	— to investigate the feasibility of performing ultra-deep sequencing of ctDNA in patients with advanced solid tumors who are currently being treated with immune checkpoint inhibitors (ICIs) — To obtain fresh tumor biopsies and serial blood samples to investigate the clonal evolution of tumors under the selection pressure of ICIs	NCT02724488
PLATON – Platform for Analyzing Targetable Tumor Mutations (Pilot-study) (PLATON)	Different cancer of the gastrointestinal tract	ctDNA, bTMB	Immune checkpoint inhibitors	— to assess genomic profiling in gastrointestinal cancer therapy and the frequencies of targetable mutations including TMB and MSI, performing NGS using the Foundation Medicine assays on tumor specimens and EDTA-whole blood samples	NCT04484636

CTCs enrichment [32−34], in 16 metastatic breast cancer patients. In 70% of analyzed patients (11/16) a subpopulation of PD-L1 expressing CTCs was detected, thus proving the hypothesis that spread metastatic circulating cells express immune checkpoint proteins. Afterward, PD-L1 expression on CTCs has been investigated in different solid tumors (breast, lung, bladder, colon, prostatic carcinoma, and HNSCC).

Most of the published studies are focused on lung cancer, in particular breast, since immunotherapy is widely used in this setting. As previously mentioned, LB can be used to obtain predictive information concerning treatment response or resistance onset as well as prognostic information. In NSCLC the presence of a higher pre-treatment PD-L1$^+$ CTC number is associated with treatment failure and defines the "non-responders" group, with a PFS lower than 6 months. Moreover, it was shown that PD-L1$^+$ CTC could be detected in all patients at progression [35]. Accordingly, it has been demonstrated that the assessment of PD-L1 status on CTCs at baseline, at 3 and 6 months after starting therapy could be used for patients' stratification; in detail, the detection of PD-L1$^+$ CTC at baseline and 3 months of treatment is associated with a poor outcome [36]. Interestingly after 6 months of treatment CTCs were detectable in all patients, PD-L1 expression on their surface could be used to dichotomize patients into two groups. Indeed, patients with PD-L1 negative CTCs all obtained clinical benefits, whereas patients with PD-L1$^+$ CTC all experienced progressive disease. These results suggest that CTCs evaluation could be used to predict patients' outcomes but also that the persistence of PD-L1 expression on CTCs might mirror a mechanism of therapy escape [36]. Nevertheless, there are still some contradictory results concerning the concordance between tissue and CTCs PD-L1 expression, indeed some papers report no correlation [35,37,38], while some others have demonstrated a concordance that can range from 57 to 100% [24,39,40]. One reason that could explain these discrepancies is the different methods used for CTCs isolation.

CTCs detection remains very challenging especially because they are very rare and are a heterogenous population. Indeed, different enrichment methods could isolate different CTCs populations, which will in turn affect PD-L1 status. It is therefore important to consider that studies might not always be comparable, especially when antigen-dependent and antigen-independent methods are used [41]. Therefore, there is an urgent need to develop clinical trials addressing two main questions: to identify the best CTCs isolation methods for the subsequent PD-L1 analysis and to evaluate the real clinical utility of PD-L1 assessment on CTCs.

5.1.3 Cell-free DNA and circulating tumor DNA

During the process of cell death (both by necrosis or apoptosis) fragments of cellular DNA are released into the extracellular environment and consequently into the bloodstream, thus being defined as ccfDNA or simply cell-free DNA (cfDNA). cfDNA is highly fragmented, fragments length is typically 180−200 base pairs, suggesting that apoptosis likely produces the majority of cfDNA in circulation [42−44]. In cancer patients, a portion of the cfDNA is released by the tumor and is, therefore, more properly defined as ctDNA. ctDNA can be easily isolated from plasma or other body fluids and can be used for several downstream molecular biology applications, including real-time PCR, droplet digital PCR (ddPCR), and Next Generation Sequencing (NGS). The amount of ctDNA can range from <0.01% to >90% of an individual's total cfDNA depending on several parameters, such as tumor burden, proliferation rate, and disease stage [45].

ctDNA has been already approved as a LB in NSCLC patients in the context of targeted therapies. Indeed, both European Medicines Agency and FDA, have approved the use of ctDNA in metastatic NSCLC for the non-invasive evaluation of *EGFR* mutational status that guides therapeutic choice [22]. Many other clinical applications have been suggested, including the possibility to use ctDNA for minimal residual disease (MRD) evaluation, early assessment of treatment response as well as early characterization of resistance mechanisms onset [46−48]. The new challenge is now to apply ctDNA in patients receiving ICIs in both adjuvant and metastatic settings. ctDNA could be used for minimal detection of predictive biomarkers, such as MSI (microsatellite instability) and TMB (tumor mutational burden) which are known to positively impact ICI responsiveness. Moreover, dynamic modifications of ctDNA during ICI could be used to monitor treatment efficacy and for the early detection of resistance, avoiding prolonged administration of ineffective treatments.

In the adjuvant setting, ctDNA could be used to detect MRD, which corresponds to the presence of residual cancer cells after treatment with local therapy. MRD evaluation could be helpful to find the perfect timing for ICI administration in an adjuvant setting. It has been shown that ICI responsiveness is inversely correlated with tumor burden [49,50], thus suggesting that the smallest the tumor the greater ICI efficacy. Consequently, the detection of MRD could indicate the smallest tumor burden and therefore the perfect time frame during which ICI could be tested as a form of early salvage therapy in patients who are about to develop disease recurrence [51]. The c-TRAK-TN trial has been designed to test this hypothesis in early-stage triple-negative breast cancer (TNBC) patients. Enrolled TNBC patients will be subjected to serial ctDNA analysis every 3 months after completion of primary treatment (either surgery, radiotherapy, or adjuvant chemotherapy). Those patients in which a rise in ctDNA level is detected, indicating a possible disease recurrence, will be randomized in a 2:1 ratio to receive the pembrolizumab as single-agent therapy or to an observation arm until disease recurrence is detected using standard modalities.

In the metastatic setting, numerous data suggest that the ctDNA level at baseline is a prognostic factor in patients receiving chemotherapy [52,53]. Based on this evidence, the same hypothesis has been examined by our group in NSCLC patients receiving nivolumab [54]. We wanted to investigate whether early dynamic changes of cfDNA in association with neutrophil to lymphocyte ratio (NLR) could predict treatment effectiveness. In our work we demonstrated that a combined increase of cfDNA and NLR >20% is associated with significantly worse survival outcomes as compared with the remaining population [54], thus providing evidence for the potential use of ctDNA as a prognostic marker. Accordingly, Goldberg et al., have shown that a drop in ctDNA level is an early marker of therapeutic efficacy and predicts prolonged survival in patients treated with ICI for NSCLC [55]. Similar results have been obtained in metastatic melanoma; patients with persistently elevated ctDNA on therapy had a poor prognosis [56−58]. ctDNA could also be used to monitor response to ICI administration. With this aim the group of Jensen et. al developed a new approach, combining cfDNA concentration and genome instability number (GIN), a metric where a higher GIN indicates more copy-number alterations. They showed that GIN can be used to discriminate clinical response from progression, differentiate progression from pseudoprogression, and identify hyper progressive disease, thus increasing the interest in the application of ctDNA testing as a surrogate biomarker of treatment response [59].

As previously mentioned MSI and TMB are becoming important biomarkers of ICI efficacy in clinical practice. MSI is due to deficiencies in DNA mismatch repair (MMR). Dysfunctional MMR is caused by several alterations including inactivating mutations in genes involved in MMR pathways (such as *MLH1*, *MSH2*, *MSH6*, or *PMS2*) or by *MLH1* promoter hypermethylation [60]. MMR impairment determines an abnormal expansion or contraction in the number of microsatellite repeats, leading to the development of tumors showing an MSI-high (MSI-H) phenotype. Mismatch-deficient tumors accumulate multiple somatic mutations which determine the formation of mutation-associated neoantigens that can stimulate the immune system [61,62]. Based on this evidence, in 2015 Le et al. conducted a phase 2 study to evaluate the clinical activity of pembrolizumab in 41 patients with progressive metastatic carcinoma with or without mismatch-repair deficiency [63]. In the study was confirmed that MMR status predicts clinical benefit to pembrolizumab, thus opening the way for the use of MSI as a predictive biomarker. These first findings were further confirmed in an expanded study and pembrolizumab was granted accelerated approval for the treatment of adult and pediatric patients with unresectable or metastatic MSI-H or MMR-deficient solid tumors [64]. Currently, MMR status is assessed by determining the presence of MSI and/or abnormal expression of MMR-related proteins in tumor tissue [65]. However, tissue may not always be available, and/or it might not be suitable for assessing MMR status; therefore, also in this context, there might be room for the use of ctDNA. PCR and NGS-based techniques can both be used for the direct detection of a change in microsatellite length; however, the PCR-based approach has one main drawback due to their sensitivity limits. Several attempts have been done to overcome this problem using other, more sensitive, techniques such as ddPCR [66]. NGS-based approaches can also be used for MSI detection in ctDNA; nevertheless, it requires high-depth sequencing combined with an optimized workflow and is still far from being routinely applied in clinical practice.

TMB is defined as the number of non-synonymous mutations present in the genome of a single tumor and it mirrors tumor genomic instability. As well as MSI, also high TMB (defined according to various cut-offs ranging from ≥ 5 to ≥ 20 mutations per Mb) has the potential to induce neoantigen production thus leading to tumor immunogenicity improvement [67]. TMB can be investigated using exclusively NGS-based approaches. TMB evaluation was initially calculated using whole-exome sequencing (WES) analysis which allows sequencing of all the coding DNA regions (exons), accounting for approximately 1% of the complete DNA sequence. Despite WES has been validated as a predictive biomarker of ICI efficacy in several tumor types (REF), it remains a challenging and expensive approach that cannot be easily widespread in laboratories. Moreover, most of the papers are focused on TMB evaluated on tumor tissue whereas despite being feasible, WES on ctDNA could be even more challenging and too costly to be used routinely in clinical practice. Therefore, researchers have attempted to develop a targeted NGS panel that could be applied also for blood-based TMB (bTMB). Zhang et al. have recently published data resulting from the assessment of TMB using an NGS Cancer Gene Panel (CGP) both on tissue and LB [68]. The aim was to explore the optimal gene panel size and algorithm to design a CGP for TMB estimation, evaluate the panel reliability, and further validate the feasibility of bTMB as a clinical actionable biomarker for immunotherapy. The study showed that the targeted-based NGS approach is cost-effective and reliable for bTMB; it was indeed reported a good correlation between bTMB calculated with CGP and tissue TMB evaluated through WES. Moreover, the authors showed that a bTMB of six or higher is positively associated with the clinical benefits of anti-PD-1 and anti-PD-L1 therapy in patients with advanced NSCLC. Stronger evidence suggesting the correlation of bTMB and ICI efficacy have been recently reported. Gandara et al. [69], have indeed analyzed bTMB in plasma samples retrieved from the OAK and POLAR trials, both designed to prove the superior efficacy of atezolizumab (anti-PD-L1 antibody) over docetaxel in NSCLC patients. Authors reported that the beneficial effects of atezolizumab (compared with docetaxel) were mostly observed in tumors with a high bTMB and that bTMB did not correlate with PD-L1 expression evaluated with IHC. Albeit being very promising, bTMB is still far from the clinic because of the lack of robust and validated tests for its routine evaluation. Translational research in oncology got us used to epochal changes and nothing suggests that this will not be the case also in the field of LB and immunotherapy.

5.1.4 Exosomes

Exosomes are biologically active lipid-bilayer extracellular vesicles with a size ranging from 30 to 120 nm and are actively secreted into the extracellular space by both normal and diseased cells, including tumor cells, through endosomal pathways. Exosomes can be considered as "cellular biopsy," indeed during their biogenesis, they engulf several biologically active molecules (mRNA, miRNA, proteins, DNA) which are characteristic of the cell from which they originate. Moreover, it has been shown that exosomes act as a cargo of information between local and distant sites, allowing intercellular communication [70,71]. It is therefore clear that exosomes participate in several physiological and pathological processes, especially in the development of cancers. As previously mentioned, exosomes can be used as LB and they seem to be involved in the immune response [72,73].

The role of exosomes in shaping the immune environment in tumors is an emerging field of interest. It has been shown that exosome-derived signals can act both as promoters and suppressors of immune responses in cancer [74]. In 1996 was first reported that exosomes derived from human lymphocyte B use major histocompatibility complex class II for antigen presentation to CD4 T cells *in vitro* [75]. More recently Hoshino et al. reported that exosomes contribute to and guide the metastatic colonization in secondary organs. This process seems to be guided by specific integrins expressed on the exosome surface that are able to dictate exosome cellular uptake. Overall, these signals induce a pro-migratory and pro-inflammatory environment, leading to bone marrow-derived myeloid cells recruitment that further stimulates inflammation [76].

As previously mentioned, tumor-derived exosomes (TEXs) and immune cell-derived exosomes are able to activate immune responses. Indeed, exosomes take part in the process of antigen transfer to APCs (antigen-presenting cells) resulting in activation of CD4 + and CD8 + T cells leading to antitumor responses enhancement and tumor progression inhibition [77]. TEXs express several tumor-associated antigens and could therefore be used as an antigen cargo in order to stimulate the immune response against tumors [78–80]. Nevertheless, there is a growing body of studies suggesting that TEXs induce tumor progression. Therefore, there is a need to deepen our knowledge concerning the role of exosomes in the immune response [81].

What seems to be clearer is the possible role of TEXs as a biomarker. Indeed, emerging evidence has shown that TEXs carry bioactive PD-L1 on their surface and can suppress the immune response [82]. In pancreatic ductal adenocarcinoma, it was shown the detection of PD-L1 + exosomes in serum correlates with poor prognosis [83]. Similar results were obtained in head and neck cancer and melanoma patients, where the prognosis is negatively influenced by levels of circulating exosomal PD-L1 [84]. Moreover, the levels of circulating exosomal PD-L1 before treatment initiation are able to discriminate between responders and nonresponders in patients treated with anti-PD1 antibodies. Indeed, nonresponders showed higher exosomal PD-L1 compared to responders, hence suggesting that exosomes could be used as a biomarker to predict the clinical benefit of immunotherapy.

As previously mentioned, exosomes carry mRNA that can be shuttled and engulfed by target cells where it can be translated into a biologically active protein. Therefore, despite the majority of papers being focused on PD-L1 protein expression, interesting data are emerging concerning the evaluation of PD-L1 mRNA inside the exosome. In a recent study, Del Re et al. investigated the expression of PD-L1 mRNA in melanoma and NSCLC patients and whether they can be used to monitor response to nivolumab and pembrolizumab. Authors showed that dynamic measurement of PD-L1 expression in plasma-derived exosomes is feasible and may provide useful information on the response to treatment with anti-PD-1 antibodies [85].

5.1.5 T-lymphocytes (the T-cell receptor) and cytokines

Despite the aforementioned and most famous LB components, there is growing interest in new circulating biomarkers. In particular, T-lymphocytes, even if not properly referable to the canonical concept of LB, have acquired a growing interest in their potentiality as a target in IO. Indeed, given the ease of manipulation and isolation of T cells from whole blood by density centrifugation, they can be specifically sorted, through flow cytometry, by selecting specific surface markers such as CD4 or CD8 [86]. In fact, both regulatory CD4 + and killing CD8 + T-cells have a pivotal role in immune surveillance against cancer cells within and beyond the tumor microenvironment; their action is generally counteracted by overexpression of surface check point proteins, such as program death ligand-1 (PD-L1), and could be re-established by the immune-restoring effects of PD-1 and PD-L1 inhibitors [87]. Interestingly, circulating CD8 + T cells overexpressing PD1 +, even if accounting for less than 5% of all peripheral CD8 + lymphocytes, demonstrated an intriguing antitumor effect by targeting tumor neo-antigens in melanoma patients [88]. These findings strongly support the idea of using circulating CD8 + /PD-1 + cells as a personalized T cell-based therapeutic approach. Furthermore, in localized clear cell renal cell carcinoma (ccRCC) the identification of an autologous characteristic peripheral blood T-cell immunophenotype helped to predict early relapse in a cohort of 40 patients [89]. Additionally, the TCR repertoire, defined as the number of T-cell clonotypes showing different TCRs, has recently emerged as a novel biomarker with intriguing possible applications in IO [90]. Indeed, each T-cell clonotype is characterized by a specific TCR which basically consists of one alpha and one beta chain each of which contains three highly variable complementarity determining regions (CDR). Of these three hypervariable regions, the variability of the CDR1 and CDR2 regions is mainly linked to germline genetic variants, otherwise, the variability of the CDR3 region is of greater interest, as a product of genetic recombination mechanisms (VDJ joining) in somatic cells. VDJ recombination is a characteristic genetic recombination process occurring in immature lymphocytes during their differentiation in the primary lymphoid organs. In detail, the gene segments (called V, D, and J) which constitute the receptors for the antigens, antibodies/immunoglobulins for B lymphocytes, and TCRs for T lymphocytes, are recombined. This process is responsible for the heterogeneity of antigen receptors for B lymphocytes and T lymphocytes [91]. Therefore, this feature gives to the CDR3 region and consequently to each TCR the uniqueness that is the basis for the determination of the TCR repertoire [92,93]. Furthermore, in the last few years, the introduction of NGS in clinical practice has allowed the identification of a TCR repertoire through the evaluation of the variability of the complementary determining region 3. Moreover, along with the TCR repertoire, NGS-based approaches allowed us to identify also the so-called TCR convergence, defined as the ratio of expanded T cell clones sharing antigen specificity but differing for CDR3 amino acid or nucleotide sequence [94]. Indeed, one of the advantages arising from the use of TCR convergence would be its ability to identify expanded T cell clones responding to tumor neoantigens stimulation. As recently reported by Looney et al., TCR convergence seems to be a feasible tool in discriminating with high accuracy (AUC = 0.77), responders from non-responders to ICI if compared to the reliability of TMB performance in the same clinical setting [94]. The use of the proper NGS platform is still under examination; in fact, given the lower error rate substitution of the Ion Torrent-based approach, if compared to Illumina-based assays, the former seemed to be more accurate and suitable for the determination of the TCR repertoire and convergence in many cohorts of cancer patients receiving anti-CTLA-4 or anti-PD-1 [94]. Several studies are still needed to understand the exact mechanism of anti-CTLA4 and anti-PD1 function, and drug responses to both seem to be regulated in different ways. Moreover, knowing the exact mechanisms through which an immunogenic tumor responds or escapes to a specific immunotherapy treatment would allow the implementation of more targeted and personalized strategies. For this purpose, the evaluation of the TCR repertoire could help in stratifying patients likely to benefit or not from anti-CTLA4 or anti-PD1 monotherapies or combinatorial therapies. Recent findings suggest that the TCR repertoire before any treatment at baseline could guide the therapeutic strategy against cancer. In particular, a study conducted in 2018 by Hogan et al. using a multi-N-plex PCR assay on the CDR3 region of the TCR, showed a higher PFS in melanoma patients treated with anti-PD-L1 inhibitors by highlighting a more clonal TCR repertoire before the treatment starting [95]. Conversely, a shorter PFS was noticed in melanoma patients, with a similar TCR repertoire shape, but receiving anti-CTLA4 inhibitors [41,96]. In detail, in the retrospective study the authors aimed to determine the performance of TCR repertoire diversity at baseline, expressed as diversity evenness index (DE50), as a predictor of anti-PD1 and anti-CTLA4 efficacy. Indeed, a clonal TCR repertoire at baseline with a DE50 < 20.4% prior to any anti-PD1 is a reliable predictor of response treatment at 12 weeks by showing a strong correlation not only with PFS but also with OS at 24 months [97,98]. Conversely, a TCR repertoire with a DE50 lower than 20.03% evaluated at baseline before any anti-CTLA4 administration is significantly associated with poor response and outcome to that specific treatment [96]. Generally, a clonally expanded TCR repertoire prevents anti-CTLA4 efficacy, above all in tumors with high PD-L1 expression, favoring a response to anti-PD(L)1 inhibitor [99]. Therefore, the shape of TCR repertoire, in terms of clonality, at baseline and throughout therapy administration seems to have predictive abilities in guiding clinical outcomes as well as response to different treatment strategies. Interestingly, several studies report that the evolution of TCR repertoire during disease is strictly related to the pressure exerted by different ICI treatments [100,101]. Indeed, the maintenance of baseline high-frequency TCR clonotypes during treatment is associated with improved clinical outcomes in melanoma patients under anti-CTLA4 Ipilimumab treatment [102]. Furthermore, in advanced NSCLC patients, a higher TCR diversity and clonality levels in circulating PD-1 + CD8 + T cells are predictive of treatment response to immune checkpoint blockade and survival outcomes before and after treatment with anti-PD1 Nivolumab, if compared with patients with low diversity [103]. In Subudhi et al. work, the authors revealed, through sequencing of the TCR β-chains, that expansion of blood CD8 T-cell clones anticipated the onset of grade 2−3 irAEs in metastatic prostate cancer patients under androgen deprivation therapy in combination with ipilimumab [104,105]. In summary, the analysis of TCR repertoire could represent a feasible tool to optimize patient selection and stratification in the context of ICI administration in several immunogenic cancers [106,107].

Circulating cytokines, such as pro-inflammatory cytokines, represent a helpful tool to monitor the dynamic behavior of the immune response against several immunogenic cancers. In the work published by Sanmamed et al., serum level of interleukin-8 (IL8) demonstrated interesting feasibility in monitoring and predicting clinical benefits in melanoma and NSCLC patients treated with ICIs in monotherapies or combinatorial regimens. In particular, IL-8 level variations showed with high specificity and sensitivity a correlation with response to anti-PD-1 in melanoma and NSCLC and in metastatic melanoma under combinatorial anti-PD-1 plus anti-CTLA-4 inhibitors. Moreover, fluctuations in IL-8 serum levels as early as 2−3 weeks after treatment initiation showed interesting reliability in predicting subsequent clinical responses. IL-8 demonstrated, also, a wide versatility in identifying pseudoprogression [108]. In 2018, Yam et al., evaluated in a cohort of melanoma patients under PD-1/PD-L1 inhibitors in monotherapy or in a combinatorial therapeutic regimen with anti-CTLA-4, the predictive role of a plasma-based cytokines assay consisting of 65 circulating cytokines. Interestingly, the identification of a panel consisting of 11 upregulated cytokines (IL13, IL2, IL1a, IL1B, IL1RA, IL12p70, IFNα2 as well as G-CSF, Fractalkine, GM-CSF, FGF-2) allowed the authors to establish the CYTOX score able to predict immune-related adverse events with high statistical power both at baseline and early during treatment [109,110].

5.1.6 The soluble form of immune-checkpoints and other circulating proteins

Physiologically, immune checkpoints maintain self-tolerance and regulate physiological immune balance by protecting healthy tissues from immune system attacks. When activated following a stimulus, T cells express PD-1 which allows them to recognize abnormal and cancerous cells. Conversely, in order to evade recognition and elimination by T cells, cancer cells express PD-L1 which binds PD-1 on T cells, rendering them inactive. Thus, the blockade of PD-1 or PD-L1 enables T-cell-mediated cancer cell death [111].

PD-1 is expressed on activated T cells, B cells, dendritic cells, monocytes, regulatory T cells, and natural killer T cells. PD-1 expression can be induced by TCRTCR-mediated activation and stimulation by cytokines such as interleukin (IL)-2, IL-7, IL-15, and IL-21 [112]. In many types of cancers, PD-1 is expressed on a large proportion of tumor-infiltrating lymphocytes (TILs) [113].

PD-L1 is an inhibitory receptor expressed on the surface of tumor cells and it is commonly upregulated in tumor cells. PD-L1 interaction with its ligand PD-1 on activated T lymphocyte inhibits its cytolytic effector functions. Tumors can create an immunosuppressive microenvironment through the overexpression of PD-L1 on tumor cells, which facilitates cancer immune evasion through the downregulation of cytotoxic T-cell activity. The blockade of the PD1/PD-L1 axis with the specific antibody inhibitors prevents T-cell suppression and promotes the immune killing of the cancer cells [113,114].

It has been hypothesized that the expression of PD-1/PD-L1 could have a predictive and/or prognostic role. Until now, the only predictive biomarker used as a complementary diagnostic test for first-line immunotherapy is PD-L1 expression assessed by IHC staining in FFPE from tissue sections. Patients with tissue PD-L1 overexpression have better responses to anti-PD-L1-directed therapy. For example, melanoma patients have a response rate of 44%−51% to anti-PD-1-directed therapy while low tissue PD-L1 expression correlates with response rates of around 6%−17%. Similarly, NSCLC patients with IHC PD-L1 overexpression have a response rate of 67%−100% while for PD-L1-negative NSCLC, the response rate was around 0%−15% [115,116].

However, PD-L1 status on tumor tissue has proven to be an imperfect predictive biomarker due to several biological and technical limitations concerning tissue sampling, intra-, and inter-tumor heterogeneity as well as diagnostic approach by IHC [41,113,117].

In addition, the expression of ICIs on immune and tumor cells is a dynamic process. Indeed, PD-L1 and PD-1 are dynamic molecules and their tissue expression, changing during disease progression and treatment, may not provide an exhaustive overview of the disease [118].

Therefore, PD-L1 status may be underestimated in small biopsies (such as bronchial and transthoracic), which are not representative of the whole tumor [41]. In this vein, the implementation of new, cost-effective, and reliable circulating biomarkers has been eagerly awaited.

Several pieces of evidence have shown that high expression of soluble PD-1 and PD-L1 (sPD-1 and sPD-L1) correlates with an unfavorable clinical outcome in different types of solid tumors [118]. The prognostic and predictive role of sPD-1 and sPD-L1 seems to be dependent on the type of cancer, leading to a different clinical response.

sPD-1 and sPD-L1 have been shown to negatively correlate with survival in NSCLC. The clinical outcome of nivolumab treatment was significantly associated with baseline plasma levels of sPD-L1 [119]. In patients with metastatic melanoma, the level of circulating exosomal PD-L1 changes during treatment with the anti-PD-1 pembrolizumab [120]; therefore, the plasma-PD-1 seems to predict the presence and the efficiency of TILs in metastatic melanoma [121]. Higher levels prior to treatment are associated with worse clinical outcomes, and increased levels during the early stages of therapy identify clinical responders [122]. High plasma levels of specific immuno-checkpoints have been shown to correlate with a dramatically poor outcome and may be used as prognostic factors in unresectable pancreatic adenocarcinoma [123]. Finally, recent evidence showed that baseline plasma levels of sPD-1, sPD-L1, and BTN3A1 predict response to nivolumab treatment in patients with metastatic RCC [124].

Hence, the actual effort is to identify peripheral blood biomarkers that reveal the dynamic and complex nature of the immune response to properly select patients who will benefit from immunotherapy [116,125].

Abbreviations

bTMB	blood tumor mutation burden
CRC	colorectal cancer
DLBCL	Diffuse large B-cell lymphomas
HNSCC	head and neck squamous cell carcinomas
MBC	metastatic breast cancer
mCRPC	metastatic castration-resistant prostate cancer
MDSC	Myeloid-Derived Suppressor Cells
NSCLC	non-small cell lung cancer
SCLC	small cell lung cancer
TNBC	triple-negative breast cancer

Key points

- ICIs have become a promising frontier for the treatment of several solid tumors, even if the treatment efficacy is variable.
- LB uses primarily blood, but also other biological fluids (plasma, serum, saliva, urine, and effusion fluids) to obtain tumor-derived material that can be used for biological testing.

- The advantage of LB over traditional solid tumor biopsy is that it is a minimally invasive, easily repeatable, time-effective procedure that better accounts for tumor heterogeneity and real-time changes in tumor dynamics.
- Using LB we can collect CTC, circulating nucleic acids such as ccfDNA, or more specifically ctDNA but also circulating vesicles such as exosomes.
- Several components, using LB, are under-investigation as predictive biomarkers of immunotherapy response or resistance, such as CTCs, cfDNA, and Exosomes, but also T-lymphocytes (the TCR), Cytokines, soluble form of Immune-Checkpoints and other circulating proteins.
- Until now, the only predictive biomarker used as a complementary diagnostic test for first-line immunotherapy is PD-L1 expression assessed by IHC staining in FFPE from tissue sections.
- LB utility in a clinical setting has been, so far, mainly investigated in the context of targeted therapies, especially in thoracic oncology. Indeed, in Advanced Non-Small Cell Lung Cancer the analysis of *EGFR* mutational status can be conducted on LB; clinicians can therefore use LB as a decision-making strategy for treatment choice.

Expert opinion

In the era of precision oncology, the introduction of LB has become a promising frontier. Non-invasive evaluation of tumor-derived biomarkers (including CTCs, ctDNA, and exosomes) isolated from peripheral blood has proven to be a viable alternative to tissue-based genotyping that can provide a comprehensive, real-time picture of tumor status. Despite recent technological and molecular advances and the many advantages of LB, to date, it is not yet considered a standard testing procedure. Further studies are needed to evaluate the accuracy of the procedure and its ability to identify various tumor types, as well as a better understanding of the technical aspects of LB detection and analysis and more efforts to standardize preanalytical and analytical procedures.

Acknowledgments

M. La Mantia, C. Brando, and V. Gristina contributed to the current work under the Doctoral Program in Experimental Oncology and Surgery, University of Palermo.

References

[1] Marin-Acevedo JA, Soyano AE, Dholaria B, Knutson KL, Lou Y. Cancer immuno-therapy beyond immune checkpoint inhibitors. J Hematol Oncol 2018;11(1):8.

[2] Yang F, Shi K, Jia YP, Hao Y, Peng JR, Qian ZY. Advanced biomaterials for cancer immunotherapy. Acta Pharmacol Sin 2020;41(7):911−27.

[3] Franklin C, Livingstone E, Roesch A, Schilling B, Schadendorf D. Immunotherapy in melanoma: recent advances and future directions. Eur J Surg Oncol 2017;43 (3):604−11.

[4] Larkin J, Chiarion-Sileni V, Gonzalez R, Grob JJ, Cowey CL, Lao CD, et al. Combined Nivolumab and ipilimumab or monotherapy in untreated melanoma. N Engl J Med 2015;373(1):23−34.

[5] Postow MA, Chesney J, Pavlick AC, Robert C, Grossmann K, McDermott D, et al. Nivolumab and ipilimumab vs ipilimumab in untreated melanoma. N Engl J Med 2015;372(21):2006−17.

[6] Gandhi L, Rodríguez-Abreu D, Gadgeel S, Esteban E, Felip E, De Angelis F, et al. Pembrolizumab plus chemotherapy in metastatic non-small-cell lung cancer. N Engl J Med 2018;378(22):2078−92.

[7] Rittmeyer A, Barlesi F, Waterkamp D, Park K, Ciardiello F, von Pawel J, et al. Atezolizumab vs docetaxel in patients with previously treated non-small-cell lung cancer (OAK): a phase 3, open-label, multicentre randomised controlled trial. Lancet. 2017;389(10066):255−65.

[8] Ravi P, Mantia C, Su C, Sorenson K, Elhag D, Rathi N, et al. Evaluation of the safety and efficacy of immunotherapy rechallenge in patients with renal cell carci-noma. JAMA Oncol 2020;6(10):1606−10.

[9] Horn L, Mansfield AS, Szczęsna A, Havel L, Krzakowski M, Hochmair MJ, et al. First-line atezolizumab plus chemotherapy in extensive-stage small-cell lung cancer. N Engl J Med 2018;379(23):2220−9.

[10] Schmid P, Adams S, Rugo HS, Schneeweiss A, Barrios CH, Iwata H, et al. Atezolizumab and nab-paclitaxel in advanced triple-negative breast cancer. N Engl J Med 2018;379(22):2108−21.

[11] Chung HC, Ros W, Delord JP, Perets R, Italiano A, Shapira-Frommer R, et al. Efficacy and safety of pembrolizumab in previously treated advanced cervical cancer: results from the phase II KEYNOTE-158 study. J Clin Oncol 2019;37(17): 1470−8.

[12] Powles T, Durán I, van der Heijden MS, Loriot Y, Vogelzang NJ, De Giorgi U, et al. Atezolizumab vs chemotherapy in patients with platinum-treated locally advanced or metastatic urothelial carcinoma (IMvigor211): a multicentre, open-label, phase 3 randomised controlled trial. Lancet. 2018;391(10122):748−57.

[13] Patel MR, Ellerton J, Infante JR, Agrawal M, Gordon M, Aljumaily R, et al. Avelumab in metastatic urothelial carcinoma after platinum failure (JAVELIN Solid Tumor): pooled results from two expansion cohorts of an open-label, phase 1 trial. Lancet Oncol 2018;19(1):51−64.

[14] Powles T, O'Donnell PH, Massard C, Arkenau HT, Friedlander TW, Hoimes CJ, et al. Efficacy and safety of durvalumab in locally advanced or metastatic urothelial carcinoma: updated results from a phase 1/2 open-label study. JAMA Oncol 2017;3 (9):e172411.

[15] Ganesh K, Stadler ZK, Cercek A, Mendelsohn RB, Shia J, Segal NH, et al. Immunotherapy in colorectal cancer: rationale, challenges and potential. Nat Rev Gastroenterol Hepatol 2019;16(6):361−75.

[16] Fuchs CS, Doi T, Jang RW, Muro K, Satoh T, Machado M, et al. Safety and effi-cacy of pembrolizumab monotherapy in patients with previously treated advanced

Liquid biopsy and immunotherapy: is all that glitter gold?   111

gastric and gastroesophageal junction cancer: phase 2 clinical KEYNOTE-059 trial. JAMA Oncol 2018;4(5):e180013.

[17] Kole C, Charalampakis N, Tsakatikas S, Vailas M, Moris D, Gkotsis E, et al. Immunotherapy for hepatocellular carcinoma: a 2021 update. Cancers (Basel) 2020;12(10).

[18] Ilie M, Long-Mira E, Bence C, Butori C, Lassalle S, Bouhlel L, et al. Comparative study of the PD-L1 status between surgically resected specimens and matched biopsies of NSCLC patients reveal major discordances: a potential issue for anti-PD-L1 therapeutic strategies. Ann Oncol 2016;27(1):147—53.

[19] Liu Y, Dong Z, Jiang T, Hou L, Wu F, Gao G, et al. Heterogeneity of PD-L1 expression among the different histological components and metastatic lymph nodes in patients with resected lung adenosquamous carcinoma. Clin Lung Cancer 2018;19 (4):e421—30.

[20] Russo A, Incorvaia L, Del Re M, Malapelle U, Capoluongo E, Gristina V, et al. The molecular profiling of solid tumors by liquid biopsy: a position paper of the AIOM-SIAPEC-IAP-SIBioC-SIC-SIF Italian Scientific Societies. ESMO Open 2021;6(3):100164.

[21] Rolfo C, Castiglia M, Hong D, Alessandro R, Mertens I, Baggerman G, et al. Liquid biopsies in lung cancer: the new ambrosia of researchers. Biochim Biophys Acta 2014;1846(2):539—46.

[22] Rolfo C, Mack PC, Scagliotti GV, Baas P, Barlesi F, Bivona TG, et al. Liquid biopsy for advanced non-small cell lung cancer (NSCLC): a statement paper from the IASLC. J Thorac Oncol 2018;13(9):1248—68.

[23] Passiglia F, Rizzo S, Rolfo C, Galvano A, Bronte E, Incorvaia L, et al. Metastatic site location influences the diagnostic accuracy of ctDNA EGFR- mutation testing in NSCLC patients: a pooled analysis. Curr Cancer Drug Targets 2018;18(7):697—705.

[24] Ilié M, Szafer-Glusman E, Hofman V, Chamorey E, Lalvée S, Selva E, et al. Detection of PD-L1 in circulating tumor cells and white blood cells from patients with advanced non-small-cell lung cancer. Ann Oncol 2018;29(1):193—9.

[25] Gascoyne PR, Noshari J, Anderson TJ, Becker FF. Isolation of rare cells from cell mixtures by dielectrophoresis. Electrophoresis. 2009;30(8):1388—98.

[26] Vona G, Sabile A, Louha M, Sitruk V, Romana S, Schütze K, et al. Isolation by size of epithelial tumor cells: a new method for the immunomorphological and molecular characterization of circulatingtumor cells. Am J Pathol 2000;156(1):57—63.

[27] Zheng S, Lin HK, Lu B, Williams A, Datar R, Cote RJ, et al. 3D microfilter device for viable circulating tumor cell (CTC) enrichment from blood. Biomed Microdevices 2011;13(1):203—13.

[28] Alix-Panabières C. EPISPOT assay: detection of viable DTCs/CTCs in solid tumor patients. Recent Results Cancer Res 2012;195:69—76.

[29] Riethdorf S, Fritsche H, Müller V, Rau T, Schindlbeck C, Rack B, et al. Detection of circulating tumor cells in peripheral blood of patients with metastatic breast cancer: a validation study of the CellSearch system. Clin Cancer Res 2007;13(3):920—8.

[30] Talasaz AH, Powell AA, Huber DE, Berbee JG, Roh KH, Yu W, et al. Isolating highly enriched populations of circulating epithelial cells and other rare cells from blood using a magnetic sweeper device. Proc Natl Acad Sci U S A 2009;106 (10):3970—5.

[31] Mazel M, Jacot W, Pantel K, Bartkowiak K, Topart D, Cayrefourcq L, et al. Frequent expression of PD-L1 on circulating breast cancer cells. Mol Oncol 2015;9 (9):1773—82.

[32] Cristofanilli M, Budd GT, Ellis MJ, Stopeck A, Matera J, Miller MC, et al. Circulating tumor cells, disease progression, and survival in metastatic breast cancer. N Engl J Med 2004;351(8):781—91.

[33] Negin BP, Cohen SJ. Circulating tumor cells in colorectal cancer: past, present, and future challenges. Curr Treat Options Oncol 2010;11(1–2):1–13.

[34] Resel Folkersma L, Olivier Gómez C, San José Manso L, Veganzones de Castro S, Galante Romo I, Vidaurreta Lázaro M, et al. Immunomagnetic quantification of circulating tumoral cells in patients with prostate cancer: clinical and pathological correlation. Arch Esp Urol 2010;63(1):23–31.

[35] Guibert N, Delaunay M, Lusque A, Boubekeur N, Rouquette I, Clermont E, et al. PD-L1 expression in circulating tumor cells of advanced non-small cell lung cancer patients treated with nivolumab. Lung Cancer 2018;120:108–12.

[36] Nicolazzo C, Raimondi C, Mancini M, Caponnetto S, Gradilone A, Gandini O, et al. Monitoring PD-L1 positive circulating tumor cells in non-small cell lung cancer patients treated with the PD-1 inhibitor nivolumab. Sci Rep 2016;6:31726.

[37] Janning M, Kobus F, Babayan A, Wikman H, Velthaus JL, Bergmann S, et al. Determination of PD-L1 expression in circulating tumor cells of NSCLC patients and correlation with response to PD-1/PD-L1 inhibitors. Cancers (Basel) 2019;11 (6).

[38] Koh Y, Yagi S, Akamatsu H, Kanai K, Hayata A, Tokudome N, et al. Heterogeneous expression of programmed death receptor-ligand 1 on circulating tumor cells in patients with lung cancer. Clin Lung Cancer 2019;20(4):270–7 e1.

[39] Dhar M, Wong J, Che J, Matsumoto M, Grogan T, Elashoff D, et al. Evaluation of PD-L1 expression on vortex-isolated circulating tumor cells in metastatic lung cancer. Sci Rep 2018;8(1):2592.

[40] Chen YL, Huang WC, Lin FM, Hsieh HB, Hsieh CH, Hsieh RK, et al. Novel circulating tumor cell-based blood test for the assessment of PD-L1 protein expression in treatment-naïve, newly diagnosed patients with non-small cell lung cancer. Cancer Immunol Immunother 2019;68(7):1087–94.

[41] Hofman P, Heeke S, Alix-Panabières C, Pantel K. Liquid biopsy in the era of immuno-oncology: is it ready for prime-time use for cancer patients? Ann Oncol 2019;30(9):1448–59.

[42] Diaz LA, Bardelli A. Liquid biopsies: genotyping circulating tumor DNA. J Clin Oncol 2014;32(6):579–86.

[43] Fan HC, Blumenfeld YJ, Chitkara U, Hudgins L, Quake SR. Noninvasive diagnosis of fetal aneuploidy by shotgun sequencing DNA from maternal blood. Proc Natl Acad Sci U S A 2008;105(42):16266–71.

[44] Jahr S, Hentze H, Englisch S, Hardt D, Fackelmayer FO, Hesch RD, et al. DNA fragments in the blood plasma of cancer patients: quantitations and evidence for their origin from apoptotic and necrotic cells. Cancer Res 2001;61(4):1659–65.

[45] Diehl F, Schmidt K, Choti MA, Romans K, Goodman S, Li M, et al. Circulating mutant DNA to assess tumor dynamics. Nat Med 2008;14(9):985–90.

[46] Cabel L, Proudhon C, Mariani P, Tzanis D, Beinse G, Bieche I, et al. Circulating tumor cells and circulating tumor DNA: what surgical oncologists need to know? Eur J Surg Oncol 2017;43(5):949–62.

[47] Calandri M, Siravegna G, Yevich SM, Stranieri G, Gazzera C, Kopetz S, et al. Liquid biopsy, a paradigm shift in oncology: what interventional radiologists should know. Eur Radiol 2020;30(8):4496–503.

[48] Alix-Panabières C, Pantel K. Clinical applications of circulating tumor cells and circulating tumor DNA as liquid biopsy. Cancer Discov 2016;6(5):479–91.

[49] Huang AC, Postow MA, Orlowski RJ, Mick R, Bengsch B, Manne S, et al. T-cell invigoration to tumour burden ratio associated with anti-PD-1 response. Nature. 2017;545(7652):60–5.

[50] Joseph RW, Elassaiss-Schaap J, Kefford R, Hwu WJ, Wolchok JD, Joshua AM, et al. Baseline tumor size is an independent prognostic factor for overall survival in

patients with melanoma treated with pembrolizumab. Clin Cancer Res 2018;24 (20):4960−7.

[51] Cabel L, Proudhon C, Romano E, Girard N, Lantz O, Stern MH, et al. Clinical potential of circulating tumour DNA in patients receiving anticancer immunotherapy. Nat Rev Clin Oncol 2018;15(10):639−50.

[52] Santiago-Walker A, Gagnon R, Mazumdar J, Casey M, Long GV, Schadendorf D, et al. Correlation of BRAF mutation status in circulating-free DNA and tumor and association with clinical outcome across four BRAFi and MEKi clinical trials. Clin Cancer Res 2016;22(3):567−74.

[53] Dawson SJ, Tsui DW, Murtaza M, Biggs H, Rueda OM, Chin SF, et al. Analysis of circulating tumor DNA to monitor metastatic breast cancer. N Engl J Med 2013;368 (13):1199−209.

[54] Passiglia F, Galvano A, Castiglia M, Incorvaia L, Calò V, Listì A, et al. Monitoring blood biomarkers to predict nivolumab effectiveness in NSCLC patients. Ther Adv Med Oncol 2019;11. Available from: https://doi.org/10.1177/1758835919839928.

[55] Goldberg SB, Narayan A, Kole AJ, Decker RH, Teysir J, Carriero NJ, et al. Early assessment of lung cancer immunotherapy response via circulating tumor DNA. Clin Cancer Res 2018;24(8):1872−80.

[56] Lee JH, Long GV, Boyd S, Lo S, Menzies AM, Tembe V, et al. Circulating tumour DNA predicts response to anti-PD1 antibodies in metastatic melanoma. Ann Oncol 2017;28(5):1130−6.

[57] Gray ES, Rizos H, Reid AL, Boyd SC, Pereira MR, Lo J, et al. Circulating tumor DNA to monitor treatment response and detect acquired resistance in patients with metastatic melanoma. Oncotarget. 2015;6(39):42008−18.

[58] Seremet T, Jansen Y, Planken S, Njimi H, Delaunoy M, El Housni H, et al. Undetectable circulating tumor DNA (ctDNA) levels correlate with favorable outcome in metastatic melanoma patients treated with anti-PD1 therapy. J Transl Med 2019;17(1):303.

[59] Jensen TJ, Goodman AM, Kato S, Ellison CK, Daniels GA, Kim L, et al. Genomewide sequencing of cell-free DNA identifies copy-number alterations that can be used for monitoring response to immunotherapy in cancer patients. Mol Cancer Ther 2019;18(2):448−58.

[60] Ryan E, Sheahan K, Creavin B, Mohan HM, Winter DC. The current value of determining the mismatch repair status of colorectal cancer: a rationale for routine testing. Crit Rev Oncol Hematol 2017;116:38−57.

[61] Segal NH, Parsons DW, Peggs KS, Velculescu V, Kinzler KW, Vogelstein B, et al. Epitope landscape in breast and colorectal cancer. Cancer Res 2008;68 (3):889−92.

[62] Gubin MM, Zhang X, Schuster H, Caron E, Ward JP, Noguchi T, et al. Checkpoint blockade cancer immunotherapy targets tumour-specific mutant antigens. Nature. 2014;515(7528):577−81.

[63] Le DT, Uram JN, Wang H, Bartlett BR, Kemberling H, Eyring AD, et al. PD-1 blockade in tumors with mismatch-repair deficiency. N Engl J Med 2015;372 (26):2509−20.

[64] Le DT, Durham JN, Smith KN, Wang H, Bartlett BR, Aulakh LK, et al. Mismatch repair deficiency predicts response of solid tumors to PD-1 blockade. Science. 2017;357(6349):409−13.

[65] Buza N, Ziai J, Hui P. Mismatch repair deficiency testing in clinical practice. Expert Rev Mol Diagn 2016;16(5):591−604.

[66] Gilson P, Merlin JL, Harlé A. Detection of microsatellite instability: state of the art and future applications in circulating tumour DNA (ctDNA). Cancers (Basel) 2021;13(7).

[67] Schumacher TN, Schreiber RD. Neoantigens in cancer immunotherapy. Science. 2015;348(6230):69−74.

[68] Wang Z, Duan J, Cai S, Han M, Dong H, Zhao J, et al. Assessment of blood tumor mutational burden as a potential biomarker for immunotherapy in patients with non-small cell lung cancer with use of a next-generation sequencing cancer gene panel. JAMA Oncol 2019;5(5):696−702.

[69] Gandara DR, Paul SM, Kowanetz M, Schleifman E, Zou W, Li Y, et al. Blood-based tumor mutational burden as a predictor of clinical benefit in non-small-cell lung cancer patients treated with atezolizumab. Nat Med 2018;24(9): 1441−8.

[70] Zhang L, Yu D. Exosomes in cancer development, metastasis, and immunity. Biochim Biophys Acta Rev Cancer 2019;1871(2):455−68.

[71] Kalluri R. The biology and function of exosomes in cancer. J Clin Invest 2016; 126(4):1208−15.

[72] Clayton A, Harris CL, Court J, Mason MD, Morgan BP. Antigen-presenting cell exosomes are protected from complement-mediated lysis by expression of CD55 and CD59. Eur J Immunol 2003;33(2):522−31.

[73] Kamerkar S, LeBleu VS, Sugimoto H, Yang S, Ruivo CF, Melo SA, et al. Exosomes facilitate therapeutic targeting of oncogenic KRAS in pancreatic cancer. Nature. 2017;546(7659):498−503.

[74] Kurywchak P, Tavormina J, Kalluri R. The emerging roles of exosomes in the modulation of immune responses in cancer. Genome Med 2018;10(1):23.

[75] Raposo G, Nijman HW, Stoorvogel W, Liejendekker R, Harding CV, Melief CJ, et al. B lymphocytes secrete antigen-presenting vesicles. J Exp Med 1996;183(3): 1161−72.

[76] Hoshino A, Costa-Silva B, Shen TL, Rodrigues G, Hashimoto A, Tesic Mark M, et al. Tumour exosome integrins determine organotropic metastasis. Nature. 2015;527(7578):329−35.

[77] Greening DW, Gopal SK, Xu R, Simpson RJ, Chen W. Exosomes and their roles in immune regulation and cancer. Semin Cell Dev Biol 2015;40:72−81.

[78] Yao Y, Chen L, Wei W, Deng X, Ma L, Hao S. Tumor cell-derived exosome-targeted dendritic cells stimulate stronger CD8 + CTL responses and antitumor immunities. Biochem Biophys Res Commun 2013;436(1):60−5.

[79] Guo D, Chen Y, Wang S, Yu L, Shen Y, Zhong H, et al. Exosomes from heat-stressed tumour cells inhibit tumour growth by converting regulatory T cells to Th17 cells via IL-6. Immunology. 2018;154(1):132−43.

[80] Rao Q, Zuo B, Lu Z, Gao X, You A, Wu C, et al. Tumor-derived exosomes elicit tumor suppression in murine hepatocellular carcinoma models and humans in vitro. Hepatology. 2016;64(2):456−72.

[81] Gao L, Wang L, Dai T, Jin K, Zhang Z, Wang S, et al. Tumor-derived exosomes antagonize innate antiviral immunity. Nat Immunol 2018;19(3):233−45.

[82] Seo N, Akiyoshi K, Shiku H. Exosome-mediated regulation of tumor immunology. Cancer Sci 2018;109(10):2998−3004.

[83] Lux A, Kahlert C, Grützmann R, Pilarsky C. c-Met and PD-L1 on circulating exosomes as diagnostic and prognostic markers for pancreatic cancer. Int J Mol Sci 2019;20(13).

[84] Theodoraki MN, Yerneni SS, Hoffmann TK, Gooding WE, Whiteside TL. Clinical significance of PD-L1. Clin Cancer Res 2018;24(4):896−905.

[85] Del Re M, Marconcini R, Pasquini G, Rofi E, Vivaldi C, Bloise F, et al. PD-L1 mRNA expression in plasma-derived exosomes is associated with response to anti-PD-1 antibodies in melanoma and NSCLC. Br J Cancer 2018;118(6): 820−4.

[86] Rosati E, Dowds CM, Liaskou E, Henriksen EKK, Karlsen TH, Franke A. Overview of methodologies for T-cell receptor repertoire analysis. BMC Biotechnol 2017;17(1):61.

[87] Sau S, Iyer AK. Immunotherapy and molecular role of T-cell in PD-1 antibody treated resectable lung cancer patients. J Thorac Dis 2018;10(8):4682−5.

[88] Gros A, Parkhurst MR, Tran E, Pasetto A, Robbins PF, Ilyas S, et al. Prospective identification of neoantigen-specific lymphocytes in the peripheral blood of melanoma patients. Nat Med 2016;22(4):433−8.

[89] Giraldo NA, Becht E, Vano Y, Petitprez F, Lacroix L, Validire P, et al. Tumor-infiltrating and peripheral blood T-cell immunophenotypes predict early relapse in localized clear cell renal cell carcinoma. Clin Cancer Res 2017;23(15):4416−28.

[90] Gibney GT, Weiner LM, Atkins MB. Predictive biomarkers for checkpoint inhibitor-based immunotherapy. Lancet Oncol 2016;17(12):e542−51.

[91] Feeney AJ, Victor KD, Vu K, Nadel B, Chukwuocha RU. Influence of the V(D)J recombination mechanism on the formation of the primary T and B cell repertoires. Semin Immunol 1994;6(3):155−63.

[92] Burtrum DB, Kim S, Dudley EC, Hayday AC, Petrie HT. TCR gene recombination and alpha beta-gamma delta lineage divergence: productive TCR-beta rearrangement is neither exclusive nor preclusive of gamma delta cell development. J Immunol 1996;157(10):4293−6.

[93] Lewis SM. The mechanism of V(D)J joining: lessons from molecular, immunological, and comparative analyses. Adv Immunol 1994;56:27−150.

[94] Looney TJ, Topacio-Hall D, Lowman G, Conroy J, Morrison C, Oh D, et al. TCR convergence in individuals treated with immune checkpoint inhibition for cancer. Front Immunol 2019;10:2985.

[95] Siravegna G, Mussolino B, Vanesio T, et al. How liquid biopsies can change clinical practice in oncology. Ann Oncol 2019;30(10):1580−90.

[96] Hogan SA, Courtier A, Cheng PF, Jaberg-Bentele NF, Goldinger SM, Manuel M, et al. Peripheral blood TCR repertoire profiling may facilitate patient stratification for immunotherapy against melanoma. Cancer Immunol Res 2019;7(1):77−85.

[97] Rolfo C, Cardona AF, et al. Challenges and opportunities of cfDNA analysis implementation in clinical practice: perspective of the International Society of Liquid Biopsy (ISLB). Crit Rev Oncol 2020;102:978.

[98] Ignatiadis M, Sledge GW, Jeffrey SS. Liquid biopsy enters the clinic-implementation issues and future challenges. Nat Rev Clin Oncol 2021;18: 297−312.

[99] Blake SJ, Ching AL, Kenna TJ, Galea R, Large J, Yagita H, et al. Blockade of PD-1/PD-L1 promotes adoptive T-cell immunotherapy in a tolerogenic environment. PLoS One 2015;10(3):e0119483.

[100] Bettegowda C, Sausen M, Leary RJ, et al. Detection of circulating tumor DNA in early- and late-stage human malignancies. Sci Transl Med 2014;224ra24.

[101] Hofman P, Heeke S, Alix-Panabières C, Pantel K. Liquid biopsy in the era of immuno-oncology: is it ready for prime-time use for cancer patients? Ann Oncol 2019;30(9):1448−59.

[102] Cha E, Klinger M, Hou Y, Cummings C, Ribas A, Faham M, et al. Improved survival with T cell clonotype stability after anti-CTLA-4 treatment in cancer patients. Sci Transl Med 2014;6(238):238ra70.

[103] Han J, Duan J, Bai H, Wang Y, Wan R, Wang X, et al. TCR repertoire diversity of peripheral PD-1. Cancer Immunol Res 2020;8(1):146−54.

[104] Subudhi SK, Aparicio A, Gao J, Zurita AJ, Araujo JC, Logothetis CJ, et al. Clonal expansion of CD8 T cells in the systemic circulation precedes development of ipilimumab-induced toxicities. Proc Natl Acad Sci U S A 2016;113(42):11919−24.

[105] Aversa I, Malanga D, Fiume G, Palmieri C. Molecular T-cell repertoire analysis as source of prognostic and predictive biomarkers for checkpoint blockade immunotherapy. Int J Mol Sci 2020;21(7).

[106] Li W, Liu JB, Hou LK, et al. Liquid biopsy in lung cancer: significance in diagnostics, prediction, and treatment monitoring. Mol Cancer 2022;21(1):25.

[107] Song P, Wu LR, Yan YH, et al. Limitations and opportunities of technologies for the analysis of cell-free DNA in cancer diagnostics. Nat Biomed Eng 2022.

[108] Sanmamed MF, Perez-Gracia JL, Schalper KA, Fusco JP, Gonzalez A, Rodriguez-Ruiz ME, et al. Changes in serum interleukin-8 (IL-8) levels reflect and predict response to anti-PD-1 treatment in melanoma and non-small-cell lung cancer patients. Ann Oncol 2017;28(8):1988−95.

[109] Lim SY, Lee JH, Gide TN, Menzies AM, Guminski A, Carlino MS, et al. Circulating Cytokines predict immune-related toxicity in melanoma patients receiving anti-PD-1-based immunotherapy. Clin Cancer Res 2019;25(5):1557−63.

[110] Fattore L, Ruggiero CF, Liguoro D, Castaldo V, Catizone A, Ciliberto G, et al. The promise of liquid biopsy to predict response to immunotherapy in metastatic melanoma. Front Oncol 2021;11:645069.

[111] Riley RS, June CH, Langer R, Mitchell MJ. Delivery technologies for cancer immunotherapy. Nat Rev Drug Discov 2019;18(3):175−96.

[112] Zhu X, Lang J. Soluble PD-1 and PD-L1: predictive and prognostic significance in cancer. Oncotarget. 2017;8(57):97671−82.

[113] Jiang Y, Chen M, Nie H, Yuan Y. PD-1 and PD-L1 in cancer immunotherapy: clinical implications and future considerations. Hum Vaccin Immunother 2019; 15(5):1111−22.

[114] Wu Q, Jiang L, Li SC, He QJ, Yang B, Cao J. Small molecule inhibitors targeting the PD-1/PD-L1 signaling pathway. Acta Pharmacol Sin 2021;42(1):1−9.

[115] Incorvaia L, Fanale D, Badalamenti G, Barraco N, Bono M, Corsini LR, et al. Programmed Death Ligand 1 (PD-L1) as a predictive biomarker for pembrolizumab therapy in patients with advanced non-small-cell lung cancer (NSCLC). Adv Ther 2019;36(10):2600−17.

[116] Patel SP, Kurzrock R. PD-L1 expression as a predictive biomarker in cancer immunotherapy. Mol Cancer Ther 2015;14(4):847−56.

[117] Incorvaia L, Fanale D, Badalamenti G, Brando C, Bono M, De Luca I, et al. A "Lymphocyte MicroRNA Signature" as predictive biomarker of immunotherapy response and plasma PD-1/PD-L1 expression levels in patients with metastatic renal cell carcinoma: pointing towards epigenetic reprogramming. Cancers (Basel) 2020;12(11).

[118] Fanale D, Incorvaia L, Badalamenti G, De Luca I, Algeri L, Bonasera A, et al. Prognostic role of plasma PD-1, PD-L1, pan-BTN3As and BTN3A1 in patients affected by metastatic gastrointestinal stromal tumors: can immune checkpoints act as a sentinel for short-term survival? Cancers (Basel) 2021;13(9).

[119] Okuma Y, Wakui H, Utsumi H, Sagawa Y, Hosomi Y, Kuwano K, et al. Soluble programmed cell death ligand 1 as a novel biomarker for nivolumab therapy for non-small-cell lung cancer. Clin Lung Cancer 2018;19(5):410−17 e1.

[120] Chen G, Huang AC, Zhang W, Zhang G, Wu M, Xu W, et al. Exosomal PD-L1 contributes to immunosuppression and is associated with anti-PD-1 response. Nature. 2018;560(7718):382−6.

[121] Incorvaia L, Badalamenti G, Rinaldi G, Iovanna JL, Olive D, Swayden M, et al. Can the plasma PD-1 levels predict the presence and efficiency of tumor-infiltrating lymphocytes in patients with metastatic melanoma? Ther Adv Med Oncol 2019;11. 1758835919848872.

[122] Fan Y, Che X, Qu J, Hou K, Wen T, Li Z, et al. Exosomal PD-L1 retains immunosuppressive activity and is associated with gastric cancer prognosis. Ann Surg Oncol 2019;26(11):3745−55.

[123] Bian B, Fanale D, Dusetti N, Roque J, Pastor S, Chretien AS, et al. Prognostic significance of circulating PD-1, PD-L1, pan-BTN3As, BTN3A1 and BTLA in patients with pancreatic adenocarcinoma. Oncoimmunology. 2019;8(4):e1561120.

[124] Incorvaia L, Fanale D, Badalamenti G, Porta C, Olive D, De Luca I, et al. Baseline plasma levels of soluble PD-1, PD-L1, and BTN3A1 predict response to nivolumab treatment in patients with metastatic renal cell carcinoma: a step toward a biomarker for therapeutic decisions. Oncoimmunology. 2020;9(1):1832348.

[125] Passiglia F, Galvano A, Gristina V, et al. Is there any place for PD-1/CTLA-4 inhibitors combination in the first-line treatment of advanced NSCLC?-a trial-level *meta*-analysis in PD-L1 selected subgroups. Transl Lung Cancer Res 2021;10 (7):3106−19. Available from: https://doi.org/10.21037/tlcr-21-52.

CHAPTER 6

Which technology performs better? From sample volume to extraction and molecular profiling

E. Capoluongo[1,2], P. Pisapia[3], U. Malapelle[3] and G. Troncone[3]
[1]Department of Molecular Medicine and Medical Biotechnology, University of Naples Federico II, Naples, Italy
[2]CEINGE, Advances Biotecnologies, Naples, Italy
[3]Department of Public Health, University of Naples Federico II, Naples, Italy

SUBCHAPTER 6.1

Molecular profiling

Learning objectives

By the end of the chapter the reader will:
- Have learned the basic concepts of the main molecular biology techniques
- Have reached in-depth knowledge of sequencing (first generation and second generation), real-time PCR, and digital PCR

The clinical implementation of liquid biopsy was allowed by Food and Drug Administration (FDA) approval of circulating tumor DNA (ctDNA) in two different patient settings; the first one is in the advanced stage nonsmall cell lung cancer (NSCLC) for the Epidermal Growth Factor Receptor (*EGFR*) gene molecular assessment to select patients for EGFR tyrosine kinase inhibitors (TKIs) administration; the second setting regards the early detection of colo-rectal cancer (CRC) through the evaluation of the methylation status of septin 9 (*SEPT9*) gene [1].

As far as the predictive role is concerned, in advanced stage NSCLC patients ctDNA analysis plays a key role in two settings of patients: naïve to any treatment (basal setting) and patients who developed a resistance to first or second-generation TKI treatments (resistance setting). In the basal setting, *EGFR* mutational status analysis can be performed on ctDNA

Liquid Biopsy
DOI: https://doi.org/10.1016/B978-0-12-822703-9.00004-1

© 2023 Elsevier Inc.
All rights reserved.

extracted from plasma, to administrate first or second-generation TKIs, in patients without availability or with inadequate results on tissue samples [1]. In the resistance setting, ctDNA extracted from plasma may be a viable option to detect the *EGFR* exon 20 resistance point mutation p. T790M, to administrate third-generation TKIs [1].

Besides the predictive role for target therapy administration, ctDNA may play a pivotal role as a noninvasive screening tool for asymptomatic people. In this setting, the evaluation of the methylation status of the *SEPT9* gene in ctDNA extracted from the plasma of asymptomatic average-risk CRC individuals may be useful to screen people who are unwilling or unable to be screened by the conventional invasive methods [2].

Nevertheless, ctDNA features several challenges related to its very low half-life (about 15 min) and concentration [<0.5% of the total circulating cell-free DNA (ccfDNA)] [3]. To this end, careful attention should be paid to adequately managing blood samples in terms of collection, ctDNA extraction, and molecular analysis [4]. First of all, it is pivotal to define how many mL of blood should be collected from each patient (section 7a). Even though there is no standard volume of blood, at least two tubes containing each 10 mL of blood should be collected [4].

Another point to be considered regards the suitability of serum samples analysis in addition to plasma specimens. Several studies demonstrated the low sensitivity of serum ctDNA analysis with respect to those results obtained on plasma specimens. For this reason, plasma should be preferred over serum [4]. Of note, in a not negligible percentage (about 15%) of cases, mutations may be detected in ctDNA extracted from serum but not in plasma samples. This evidence suggests that the analysis of both samples may allow an increase in the sensitivity of ctDNA analysis [5].

Besides the volume and sample type, it is important to choose an adequate methodology for blood collection (section 7b). Different tubes are commercially available. Ethylenediaminetetra-acetic acid (EDTA)-containing tubes (Vacutainer, BD, Plymouth, UK) may avoid clotting, which can lead to cfDNA sequestration, and are less expensive than preservative tubes [PAXgene Blood DNA tubes (Qiagen, Hilden, Germany) or Cell-Free DNA BCT tubes (Streck, La Vista, NE, USA)]. However, the adoption of EDTA-containing tubes requires rapid sample processing (within 2 h) with respect to preservative tubes (up to 14 days) [6—8].

The following step is represented by centrifugation (section 7b). This phase is indispensable for ctDNA purification [1].

The last preanalytical step is represented by ctDNA extraction (section 7b). Also, in this case, a plethora of commercial kits are available. On the overall, Mauger et al. among eleven ctDNA extraction kits demonstrated a higher accuracy and reproducibility for the three following kits: QIAamp circulating nucleic acid kit (Qiagen, Hilden, Germany), the Norgen Plasma/Serum Circulating DNA Purification Mini Kit, and the Norgen Plasma/Serum Cell-Free Circulating DNA Purification Mini Kit (Norgen, Thorold, ON, Canada) [9]. In addition, another comparative study by Malapelle et al. reported that the QIA symphony DSP Virus/Pathogen Midi Kit (Qiagen) on the QIA symphony automatic platform (Qiagen) performs as good as QIAamp circulating nucleic acid with a reduction of costs [5]. More recently, novel microfluidic technologies are under investigation for ctDNA extraction [10].

Despite the central role of ctDNA in clinical practice, several other biomarkers may potentially be analyzed in liquid biopsy such as circulating tumor cells (CTCs; section 7c), circulating exosomes (section 7c), and circulating tumor RNA (ctRNA; section 7d) [11].

Another relevant step in ctDNA evaluation is represented by the molecular technique to adopt (section 7e). Two approaches may be employed: polymerase chain reaction (PCR) based technologies [such as real-time PCR (RT-PCR) or digital PCR (dPCR)] or sequencing-based platforms (next-generation sequencing [NGS]) (Fig. 6.1.1). Despite PCR-based technologies

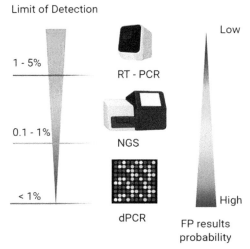

Figure 6.1.1 Schematic representation of correlation among Limit of Detection, type of technology. [Real Time PCR (RT − PCR), Next-Generation Sequencing (NGS) and digital-PCR (dPCR)] and probability of false positive (FP) results.

are generally acceptable for ctDNA analysis, due to short turnaround time [TAT], limited costs, FDA approved and employed in predictive molecular pathology laboratories for another biomarker testing, these techniques show a limited reference range (related to the adopted specific probes) and cannot identify some clinical unknown rare mutations [12,13]. This limitation may be overcome by NGS which allows the analysis of different clinical relevant hotspots and novel rare mutations within the gene panel for different patients, simultaneously [5]. However, NGS technologies require careful validation, highly trained personnel, and sophisticated bioinformatics tools, before their implementation in clinical practice [14]. Finally, attention should be paid to the standardization and quality assurance in liquid biopsy testing (section 7f).

6.1.1 Expert opinion

In the era of precision medicine, tumor molecular profiling is crucial to leading a tailored oncological treatment plan. Accordingly, molecular biology techniques have evolved into a fundamental part of the clinical oncology routine. These novel and modern techniques determine an extraordinary tool for clinical and translational research in precision oncology. Thusly, modern oncology is perfectly integrated with molecular biology; so it is essential to have a better comprehension of the main techniques to understand its advantages, but also its limits.

6.1.2 Key points

- Understand the main molecular biology techniques that are routinely used in clinical practice such as NGS and ddPCR
- Digital PCR should be considered as an upgrade real-time PCR technique and it is currently used in clinical practice
- Debate, whether a pathogenetic variant identified with 0,01% frequency, is clinically relevant and it could guide medical decision

References

[1] Pisapia P, Malapelle U, Troncone G. Liquid biopsy and lung cancer. Acta Cytol 2019;63:489−96.
[2] Church TR, Wandell M, Lofton-Day C, Mongin SJ, Burger M, Payne SR, et al. PRESEPT Clinical Study Steering Committee, Investigators and Study Team. Prospective evaluation of methylated SEPT9 in plasma for detection of asymptomatic colorectal cancer. Gut. 2014;63:317−25.

[3] Malapelle U, Pisapia P, Rocco D, Smeraglio R, di Spirito M, Bellevicine C, et al. Next generation sequencing techniques in liquid biopsy: focus on non-small cell lung cancer patients. Transl Lung Cancer Res 2016;5:505−10.

[4] Rolfo C, Mack PC, Scagliotti GV, Baas P, Barlesi F, Bivona TG, et al. Liquid biopsy for advanced non-small cell lung cancer (NSCLC): a statement paper from the IASLC. J Thorac Oncol 2018;13:1248−68.

[5] Malapelle U, Mayo de-Las-Casas C, Rocco D, Garzon M, Pisapia P, Jordana-Ariza N, et al. Development of a gene panel for next-generation sequencing of clinically relevant mutations in cell-free DNA from cancer patients. Br J Cancer 2017;116:802−10.

[6] Sherwood JL, Corcoran C, Brown H, Sharpe AD, Musilova M, Kohlmann A. Optimised pre-analytical methods improve KRAS mutation detection in circulating tumour DNA (ctDNA) from patients with non-small cell lung cancer (NSCLC). PLoS One 2016;11:e0150197.

[7] Medina Diaz I, Nocon A, Mehnert DH, Fredebohm J, Diehl F, Holtrup F. Performance of streck cfDNA blood collection tubes for liquid biopsy testing. PLoS One 2016;11:e0166354.

[8] Rothwell DG, Smith N, Morris D, Leong HS, Li Y, Hollebecque A, et al. Genetic profiling of tumours using both circulating free DNA and circulating tumour cells isolated from the same preserved whole blood sample. Mol Oncol 2016;10:566−74.

[9] Mauger F, Dulary C, Daviaud C, Deleuze JF, Tost J. Comprehensive evaluation of methods to isolate, quantify, and characterize circulating cell-free DNA from small volumes of plasma. Anal Bioanal Chem 2015;407:6873−8.

[10] Campos CDM, Gamage SST, Jackson JM, Witek MA, Park DS, Murphy MC, et al. Microfluidic-based solid phase extraction of cell free DNA. Lab Chip 2018;18:3459−70.

[11] Crowley E, Di Nicolantonio F, Loupakis F, Bardelli A. Liquid biopsy: monitoring cancer-genetics in the blood. Nat Rev Clin Oncol 2013;10:472−84.

[12] Oxnard GR, Thress KS, Alden RS, Lawrance R, Paweletz CP, Cantarini M, et al. Association between plasma genotyping and outcomes of treatment with osimertinib (AZD9291) in advanced non-small-cell lung cancer. J Clin Oncol 2016;34:3375−82.

[13] Malapelle U, Sirera R, Jantus-Lewintre E, Reclusa P, Calabuig-Fariñas S, Blasco A, et al. Profile of the roche cobas® EGFR mutation test v2 for non-small cell lung cancer. Expert Rev Mol Diagn 2017;17:209−15.

[14] Vigliar E, Malapelle U, de Luca C, Bellevicine C, Troncone G. Challenges and opportunities of next-generation sequencing: a cytopathologist's perspective. Cytopathology 2015;26:271−83.

Further reading

Russo A, Incorvaia L, Del Re M, Malapelle U, Capoluongo E, Gristina V, et al. The molecular profiling of solid tumors by liquid biopsy: a position paper of the AIOM-SIAPEC-IAP-SIBioC-SIC-SIF Italian Scientific Societies. ESMO Open 2021;6(3):100164. Available from: https://doi.org/10.1016/j.esmoop.2021.100164. Epub 2021 Jun 3. PMID: 34091263.

Russo A, Incorvaia L, Malapelle U, Del Re M, Capoluongo E, Vincenzi B, et al. The tumor-agnostic treatment for patients with solid tumors: a position paper on behalf of the AIOM- SIAPEC/IAP-SIBioC-SIF Italian Scientific Societies. Crit Rev Oncol Hematol 2021;165:103436. Available from: https://doi.org/10.1016/j.critrevonc.2021.103436. Epub 2021 Aug 8. PMID: 34371157.

SUBCHAPTER 6.2

Biological fluid withdrawal: how much sample volume is enough?

E. Capoluongo[1,2]
[1]Department of Molecular Medicine and Medical Biotechnology, University of Naples Federico II, Naples, Italy
[2]CEINGE, Advances Biotecnologies, Naples, Italy

Learning objectives

By the end of the chapter the reader will:
• understand the fundament of liquid biopsy
• have a deep knowledge of ctDNA analysis in earlier diagnosis or screening

6.2.1 Introduction

Cancer biomarkers can be detected in different bodily fluids, such as blood, urine, saliva, cerebrospinal fluid, stool, and lavage effusions (Fig. 6.2.1). Concentrations of cfDNA in blood plasma were reported

Figure 6.2.1 ctDNA quantification in biofluids samples.

as ranging from about 1.8 to 44 ng/mL with a half-life less than 2.5 h: the amount of plasma DNA is much higher in patients with cancer (5−1500 ng m/L) than in healthy individuals (1−5 ng m/L) [15], with allelic frequency (the fraction of mutated/ wild type alleles) in blood being very variable in patients with cancer [16].

The interest for liquid biopsy (LB)-based assays is rapidly increasing to detect driver mutations in such genes and to better understand mechanisms of resistance for targeted therapies in many cancers. The importance to identify and enrich specific tumor actionable mutations underlines the need to develop molecular methods which permit the detection and monitoring of multiple classes of mutation with the highest sensitivity: ctDNA could have utility at almost every stage of cancer patient management, including early diagnosis, diagnosis, minimally invasive molecular profiling, treatment monitoring, detection of residual disease, and identification of resistance mutations [17]. Unfortunately, the sensitivity of the most widely used blood-based tests is limited in specific conditions of low tumor volume where the shed of tumor-derived elements can be limited. Liquid biopsy is not limited to plasma because tumor DNA circulating (ctDNA) in other body fluids such as urine, pleural fluid, cerebrospinal fluid, or cytology specimen-derived supernatant can be analyzed. Compared to cell blocks, these fluids in tight contact with the tumor mass can contain a more abundant and less analytically demanding tumor DNA [18].

The importance of starting volume is particularly emphasized when ctDNA analysis is used as a tool for both quantitative analyses of disease burden and genomic analysis [19]. The isolation and characterization of ctDNA in individuals before a cancer diagnosis, or in pre-symptomatic subjects, opens the way for the use of ctDNA analysis in earlier diagnosis or screening [20].

Monitoring multiple mutations in parallel can enhance sensitivity for ctDNA detection, enabling the assessment of clonal evolution of a patient's disease: this allows the identification of resistance mutations before observing the clinical progression [21]. Although the analysis of gene panels in plasma has become available as a potential tool, some studies are ongoing to establish the complete performance and clinical utility of such assays when a tumor biopsy is not available for analysis. Potential applications of ctDNA were published by some proof-of-principle studies: prospective clinical trials are underway to assess the clinical utility of ctDNA analysis for molecular profiling and disease monitoring. Nevertheless, to fully exploit the potential utility of liquid biopsies-based assays, the biology of cfDNA and ctDNA must be investigated further. Mechanisms of release and degradation, and the factors that affect the representation of ctDNA in plasma, are still poorly understood. The characteristics of ctDNA will be clarified through large clinical studies where variables should be deeply controlled. In cancer patients, shorter ctDNA fragments compared to normal cfDNA fragments in the range of 90–150 bp are detected in blood plasma [21].

The ctDNA floats (alone or within exosomes) in many body fluids such as plasma, serum, urine, pleural fluid, cerebrospinal fluid, or cell supernatant: obviously, both the location of the tumor and the stage of the disease can influence the release of different amounts [22]. While in the laboratory setting, through spike-in experiments in EDTA tubes the isolation of ctDNA low amounts of tumor cells (0.1%) was obtained, it is very challenging to have a gold standard for volumes of body fluids to homogenize the ctDNA recovery. The main feature of ctDNA is that these molecules are shorter than nonmutant cfDNA in plasma, as demonstrated by PCR and sequencing methods [21,22]. While we know that some analytical issues can influence the rate of ctDNA recovery, it is not clear the impact of preanalytical factors, particularly the volume of starting biological fluid. Coming back to the analytical phase, it is reported that the commonly used library preparation methods introduce some biases: single-stranded DNA library preparation can recover DNA fragments with damaged ends, and when applied to cfDNA, it uncovered a large part of fragments shorter than 100 bases. Moreover, different extraction and sequencing methods may provide complementary data, that could be helpful, particularly when matched to those coming from tissues [22].

Tumor burden in plasma is generally assessed by quantifying mutations (or other genome alterations) that were previously detected in the patient's tumor sample. Therefore, the different technologies (such as NGS, digital droplet PCR, and shallow whole-genome sequencing) should work in a complementary way [23]. For mutation calling across a panel of genes or hotspots, the risk of false positives raises in the function of the size of the panel: bioinformatics filters are necessary to increase specificity, which reduces sensitivity for rare variants [24]. Preliminary knowledge of the mutation profile (as that coming from tumor sequencing data) enables the detection of known individual-specific mutations above the background error rate, as opposed to calling mutations de novo [23,24]. Therefore, sequencing-based assays can be used as sensitive and quantitative tools for ctDNA detection, quantitation, and monitoring, in addition to their use for mutation profiling. Since ctDNA can be quantified by different metrics, such as mutant allele concentration (copies per mL) or mutant allele fraction, these would be affected differently by analytical, preanalytical, and physiological characteristics [25]. In detail, preanalytical factors affecting the release of germline DNA from blood cells would significantly reduce the mutant allele fraction: thus, when considering the different tumor stages, where ctDNA concentration in plasma correlates with tumor size, these factors can show a dramatic impact on the analytical sensitivity of the assay [23−26]. It is reported that mutant alleles in plasma increase in their fraction by around 0.08%, and concentration by 6 mutant copies/milliliter of plasma, for every cm^3 of disease mass. For example, poor tumor vascularisation could impede or promote ctDNA release into the bloodstream by means of hypoxia and cell death. Histological differences could influence both the rate and type of cell death.

The relationship between ctDNA amounts and cancer stage is in favor of a prognostic utility for ctDNA: cancer patients with detectable ctDNA showed worse survival outcomes than those without. Serial liquid biopsies may have particular utility for adaptive or reactive therapy, where resistance-associated mutations are prospectively identified, and therapy adapted in real-time [23−25].

The impact of preanalytical factors depends on the biomarker and cavitary fluid analyzed and should be accurately determined for each newly developed liquid biopsy assay.

Among the most analyzed preanalytical variables, particularly regarding the CTC detection, the following were better investigated: (a) the type of blood collection tube, (b) time between sampling analysis and (c) storage temperature of whole blood. Depending on the downstream method of analysis, preanalytical variables may have different impacts. There are about nine different types of collection tubes specifically designed for the preservation of cell-free nucleic acids and methylated DNA. Although the National Cancer Institute Biorepositories and Biospecimen Research Branch Biospecimen Evidence-Based Practices recommend shorter durations of room temperature storage for both EDTA (2−4 h) and preservative tubes (up to 3 days), commercially available tubes used to collect blood, allow for its shipment and storage at 18°C−25°C (room temperature) for plasma processing 3 to 7 days after collection [17,27,28].

The blood volume used for molecular profiling of cfDNA represents an important issue particularly when low-abundant VAFs need to be called. A study based on modeling showed that to be able to detect de novo a single mutation with a VAF of 0.01% with 95% confidence would need 150−300 mL with 30,000 × sequencing coverage. Furthermore, it has been demonstrated that the sensitivity of a given ctDNA assay in patients with localized cancers is dependent on both the blood volume analyzed and the number of mutations screened [29]. Incoming new molecular approaches for personalized tracking of mutations, particularly the so-called targeted digital sequencing, higher blood volumes (one to two tubes of blood) are used to detect ctDNA at very low concentrations [29].

The role of quality of starting sample and volume has recently been reported by Saunders [30], underlying as plasma samples are often limited and they can show undesirable characteristics, such as lipemia or hemolysis, which contribute undesirable genomic DNA to the sample.

Finally, regardless of the extraction method, multiple freeze-thawing steps of blood plasma and freeze-thawing of whole blood are not recommended [17].

When ctDNA analysis is requested, using plasma instead of serum is preferred because the latter holds higher levels of nontumor cfDNA deriving from leukocyte lysis which occurs with blood clotting. The dilution of ctDNA by leukocyte cfDNA can adversely affect the detection of the previous, particularly ctDNA harboring low allele fraction mutations [31].

Unfortunately, few data are reported on the effects of individual-related factors, such as pregnancy, lifestyle, and some nonmalignant situations that could alter cfDNA levels in blood [32].

Although specific guidelines have been proposed concerning the preanalytical conditions for cfDNA testing, including quality control in blood collection, plasma preparation, cfDNA extraction, and quantification procedure variabilities [33] and cfDNA storage, blood sample transport conditions, and biological and demographic factors.

6.2.2 Other body fluids used in liquid biopsy-based assays

The DNA characterization in urines to explore its potential as a source of cfDNA for cancer detection was reported by Jain et al. [34]. Urines represent the truly noninvasive liquid biopsy, that allows for the easiest collection, storage, processing, and shipment when compared to the blood. The collection of urine does not give any discomfort to the patient and does not necessitate trained medical staff. Since larger collection volumes can be obtained, the diluted concentration of some ctDNA in urine can be therefore overcome. Urine also contains less contaminating proteins as compared with serum/plasma, making the process of DNA isolation easier. Not last, urine collection can be performed at home and then shipped to certified laboratories for testing facilitating any step surrounding the sample collection. The use provided by urine-based liquid biopsy will enable the regular monitoring of high-risk populations, such as those infected with both infectious, chronic, or oncological liver disorders [34]. These features increase patient compliance for cancer screening and allow more frequent monitoring to enable patients to seek curative treatment and care [35]. Nevertheless, since there are no standardized protocols or guidelines in support of the use of urines or other body fluid/effusion as complementary or alternative to plasma, the implementation of an alternative to blood-LB-based test is still far to be achieved. Among tests under evaluation for the next implementation in a clinical setting, *the Urine exosome-based genomic test* for the diagnosis of prostate cancer is within the FDA's Breakthrough Devices Program [36,37]. None of the tests running in the laboratories for translational research purposes have been implemented in the diagnostic setting.

Finally, although some studies have demonstrated several proof-of-principle concepts regarding the potential application CSF-based approach to trace tumor-derived molecules in brain tumors, there are several challenges to be overcome in the future. The improvement of signal detection along with lack of methodological standardization are still limiting the use of CSF and blood in these tumors: noteworthy, among the main limits regarding the reproducibility of biomarker studies there is a lack of guidelines and standardized protocols, with different issues relating to cohort sizes, management of confounding factors, as well as a choice of control groups [38].

Unfortunately, these issues are still affecting all applications of LB and still represent the main limits for full implementation in a clinical setting.

The volume of blood, CSF, and other body fluids represent only a part of the issues limiting the standardization of protocols: nevertheless, particularly in this setting it should be encouraged a consensus to establish: (a) how much volume of blood or other body fluid/effusion is necessary to collect for each cancer patient, possibly differentiating between early and advanced conditions;

(b) when to collect, particularly the time of sampling after the so-called "time 0"; (c) how much plasma volume must be used for the extraction.

This could be the first step towards an attempt at harmonization.

6.2.3 Key points

- The blood volume used for molecular profiling of cfDNA represents an important issue particularly when low-abundant VAFs need to be called
- Since there are no standardized protocols or guidelines in support of the use of urines or other body fluid/effusion as complementary or alternative to plasma, the implementation of an alternative to blood-LB based test is still far to be achieved
- Lack of methodological standardization is still limiting the use of CSF and blood in brain tumors
- Several issues are still affecting all applications of Liquid Biopsy and still represent the main limits for full implementation in a clinical setting.

References

[15] Fleischhacker M, Schmidt B. Circulating nucleic acids (CNAs) and cancer-a survey. Biochim Biophys Acta 2007;1775:181−232.

[16] Mouliere F, El Messaoudi S, Gongora C, Guedj AS, Robert B, Del Rio M, et al. Circulating cell-free DNA from colorectal cancer patients may reveal high KRAS or BRAF mutation load. Transl Oncol 2013;6:319−28.

[17] Agrawal L, Engel KB, Greytak SR, Moore HM. Understanding preanalytical variables and their effects on clinical biomarkers of oncology and immunotherapy. Semin Cancer Biol 2018;52(Pt 2):26−38.

[18] Geeurickx E, Hendrix A. Targets, pitfalls and reference materials for liquid biopsy tests in cancer diagnostics. Mol Asp Med 2020;72:100828. Available from: https://doi.org/10.1016/j.mam.2019.10.005.

[19] De Mattos-Arruda L, Siravegna G. How to use liquid biopsies to treat patients with cancer. ESMO Open 2021;6(2):100060. Available from: https://doi.org/10.1016/j.esmoop.2021.100060.

[20] Siravegna G, Mussolin B, Venesio T, Marsoni S, Seoane J, Dive C, et al. How liquid biopsies can change clinical practice in oncology. Ann Oncol 2019;30:1580−90.

[21] Zhang L, Liang Y, Li S, Zeng F, Meng Y, Chen Z, et al. The interplay of circulating tumor DNA and chromatin modification, therapeutic resistance, and metastasis. Mol Cancer 2019;18:36.

[22] Crowley E, Di Nicolantonio F, Loupakis F, Bardelli A. Liquid biopsy: monitoring cancer-genetics in the blood. Nat Rev Clin Oncol 2013;10:472−84.

[23] Chen M, Zhao H. Next-generation sequencing in liquid biopsy: cancer screening and early detection. Hum Genomics 2019;13:34.

[24] Wang JF, Pu X, Zhang X, Chen K, Xi Y, Wang J, et al. Variants with a low allele frequency detected in genomic DNA affect the accuracy of mutation detection in cell-free DNA by next-generation sequencing. Cancer 2018;124:1061−9.

[25] Wan JCM, Massie C, Garcia-Corbacho J, Mouliere F, Brenton JD, Caldas C, et al. Liquid biopsies come of age: towards implementation of circulating tumour DNA. Nat Rev Cancer 2017;17(4):223−38.

[26] Boonstra PA, Wind TT, van Kruchten M, Schuuring E, Hospers GAP, van der Wekken AJ, et al. Clinical utility of circulating tumor DNA as a response and follow-up marker in cancer therapy. Cancer Metastasis Rev 2020;39:999−1013.

[27] Grölz D, Hauch S, Schlumpberger M, Guenther K, Voss T, Sprenger-Haussels M, et al. Liquid biopsy preservation solutions for standardized pre-analytical workflows-venous whole blood and plasma. Curr Pathobiol Rep 2018;6:275−86.

[28] Van Paemel R, De Koker A, Caggiano C, Morlion A, Mestdagh P, De Wilde B, et al. Genome-wide study of the effect of blood collection tubes on the cell-free DNA methylome. Epigenetics 2020;1−11.

[29] McDonald BR, Contente-Cuomo T, Sammut SJ, Odenheimer-Bergman A, Ernst B, Perdigones N, et al. Personalized circulating tumor DNA analysis to detect residual disease after neoadjuvant therapy in breast cancer. Sci Transl Med 2019;11(504): eaax7392.

[30] Saunders HE. Liquid biopsy quality control and the importance of plasma quality, sample preparation, and library input for next generation sequencing analysis. J Biomol Tech 2019;30(Suppl):S26.

[31] Pittella-Silva F, Chin YM, Chan HT, Nagayama S, Miyauchi E, Low SK, et al. Plasma or serum: which is preferable for mutation detection in liquid biopsy? Clin Chem 2020;66:946−57.

[32] Merker JD, Oxnard GR, Compton C, Diehn M, Hurley P, Lazar AJ, et al. Circulating tumor DNA analysis in patients with cancer: American Society of Clinical Oncology and College of American Pathologists Joint Review. J Clin Oncol 2018;36:1631−41.

[33] Meddeb R, Pisareva E, Thierry A. Guidelines for the preanalytical conditions for analyzing circulating cell-free DNA. Clin Chem 2019;65:623−33.

[34] Jain S, Lin SY, Song W, Su YH. Urine-based liquid biopsy for nonurological cancers. Genet Test Mol Biomarkers 2019;23 277-28.

[35] Hann HW, Jain S, Park G, Steffen JD, Song W, Su YH. Detection of urine DNA markers for monitoring recurrent hepatocellular carcinoma. Hepatoma Res 2017;3:105−11.

[36] Arechederra M, Ávila MA, Berasain C. Liquid biopsy for cancer management: a revolutionary but still limited new tool for precision medicine. Adv Lab Med 2020;1(3) 20200009.

[37] Oshi M, Murthy V, Takahashi H, Huyser M, Okano M, Tokumaru Y, et al. Urine as a source of liquid biopsy for cancer. Cancers (Basel) 2021;13:2652.

[38] Shankar GM, Balaj L, Stott SL, Nahed B, Carter BS. Liquid biopsy for brain tumors. Expert Rev Mol Diagn 2017;17:943−7.

Further reading

Froelich MF, Capoluongo E, Kovacs Z, Patton SJ, Lianidou ES, Haselmann V. The value proposition of integrative diagnostics for (early) detection of cancer. On behalf of the EFLM interdisciplinary Task and Finish Group "CNAPS/CTC for early detection of cancer" [published online ahead of print, 2022 Feb 24] Clin Chem Lab Med 2022. Available from: https://doi.org/10.1515/cclm-2022-0129.

Russo A, Incorvaia L, Malapelle U, Del Re M, Capoluongo E, Vincenzi B, et al. The tumor-agnostic treatment for patients with solid tumors: a position paper on behalf of the AIOM-SIAPEC/IAP-SIBioC-SIF Italian Scientific Societies. Crit Rev Oncol Hematol 2021;165:103436. Available from: https://doi.org/10.1016/j.critrevonc.2021.103436 Epub 2021 Aug 8. PMID: 34371157.

SUBCHAPTER 6.3

Methods for cf/ct DNA isolation

U. Malapelle[1], P. Pisapia[1], A. Galvano[2], V. Gristina[2], M. La Mantia[2] and G. Troncone[1]

[1]Department of Public Health, University of Naples Federico II, Naples, Italy
[2]Department of Surgical, Oncological, and Oral Sciences, University of Palermo, Palermo, Italy

Learning objectives

By the end of the chapter, the reader will:

- Have a deep knowledge of the methods for cell-free (cf)/circulating tumor DNA (ct DNA) isolation
- Know the international guidelines and recommendations for cf/ct DNA isolation

It is widely held, that ctDNA plays a relevant diagnostic, predictive, and prognostic role in most cancer patients. However, as a requirement, this analyte should be handled with care. It has a very low half-life (about 15 min) and concentration (<0.5% of the total circulating cell-free DNA [ccfDNA]) that require very sensitive techniques that should be carefully validated and monitored [39].

For these reasons, the International Association for the Study of Lung Cancer (IASLC) established in a statement paper a few crucial recommendations. These need to be strictly followed to ensure that plasma samples are adequately handled. In particular special care should be taken in three crucial preanalytical phases: blood collection, centrifugation, and ctDNA extraction [40].

As far as blood collection is concerned, laboratories may adopt different collection tubes. In particular, ethylenediaminetetra-acetic acid (EDTA)–containing tubes (Vacutainer, BD, Plymouth, UK) are less costly than preservative tubes, also having the advantage to avoid clotting [41]. However, blood samples collected in EDTA-containing tubes should be rapidly processed (within 1−2 h) to limit ctDNA degradation [42]. To this end, when adopting these tubes, blood collection and blood analysis should be planned in a single workflow to be carried out not only, in the same Institution but ideally in the laboratory. To this end, our laboratory has staffed a research nurse to be the bridge between blood collection and nucleic acid extraction. Needless to say, this optimal approach also has the advantage to limit turnaround time [43]. Conversely, PAXgene Blood DNA tubes (Qiagen, Hilden, Germany) or Cell-Free DNA BCT tubes (Streck, La Vista, NE, USA) allow a slightly delayed analysis of ctDNA, still ensuring high-quality results [44]. These tubes contain formaldehyde-free preservative agents able to prevent leucocyte lysis and significantly reduce ctDNA degradation [45]. Interestingly, it has been demonstrated that Cell-Free DNA BCT tubes perform better than PAXgene Blood DNA tubes in preventing cell lysis [46]. As a general rule, to optimize the results, the IASLC guideline recommends that the time between blood withdrawal to ctDNA extraction should not exceed 2 h for EDTA tubes and three days for preservative tubes [40].

The second crucial step is represented by centrifugation. This phase enables more efficient purification of plasma samples, by removing the vast majority of the blood-formed elements (in particular white cells containing genomic nonneoplastic DNA) [42]. To ensure an optimal purification of plasma samples, Malapelle et al. describe two centrifugation steps (each of 2300 rpm for 10 min at room temperature) based on protocol. This procedure allowed to obtain about 2 mL of clarified plasma, starting from up to 10 mL of blood [43]. Sorber et al. reported that the adapted European Committee for Standardization (CEN) protocol (two-step, 1900 rpm for 10 min and then 16000 rpm for 10 min at room temperature) generates high-quality ctDNA [47]. In another study, Page et al. performed three centrifugation steps (the first at 1000 rpm for 10 min at 4°C, the second at 1000 rpm, or 2000 rpm, or 10000 rpm for 10 min at 4°C, and the third, performed after thawing, at 1000 rpm for 5 min at room temperature) [48]. Overall, the IASLC statement paper suggested centrifuging at least twice the blood samples (for the first time in the original tube and then, the plasma obtained should be further centrifuged in a second tube) [40].

The third crucial step in ctDNA isolation is represented by extraction. To date, several kits are commercially available. In a recent study, the accuracy and reproducibility of 11 different methods for ctDNA extraction have been compared. Overall, the authors highlighted that the better results in terms of accuracy and reproducibility of ctDNA extraction can be obtained by three kits: QIAamp circulating nucleic acid kit (Qiagen, Hilden, Germany), the Norgen Plasma/Serum Circulating DNA Purification Mini Kit, and the Norgen Plasma/Serum Cell-Free Circulating DNA Purification Mini Kit (Norgen, Thorold, ON, Canada) [49]. Similar results were reported by Sorber et al. who underlined the high efficacy of both the QIAamp circulating nucleic acid kit and of the Maxwell RSC ccfDNA Plasma Kit (Promega, Leiden, Netherlands). This latter has the additional advantage of magnetic beads—based fully automated protocol [50]. More recently it was suggested that the QIAamp circulating nucleic acid can efficiently be replaced by the QIA symphony DSP Virus/Pathogen Midi Kit on the QIA symphony automatic platform [43].

In conclusion, these above discussed preanalytical steps are key. Mutational assessments in blood or other body fluid sources of cancer patients are reliable, only when high standard technical procedures are validated, implemented, and monitored.

6.3.1 Key points

- The first essential step in ct\cf DNA isolation is blood collection
- The second crucial step is represented by centrifugation.
- The third crucial step in ctDNA isolation is represented by extraction
- Mutational assessments in blood or other body fluid sources of cancer patients are reliable, only when high standard technical procedures are validated, implemented, and monitored (Fig. 6.3.1).

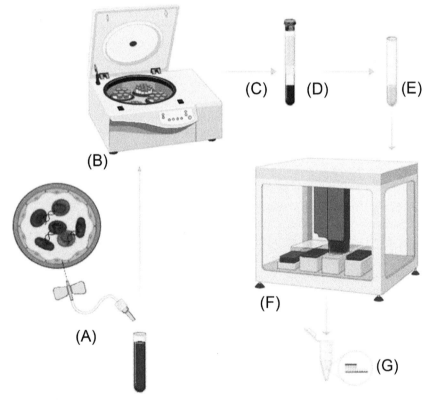

Figure 6.3.1 Schematic representation of blood collection (A) by using butterfly and dedicated tube, plasma centrifugation (B) carried out two centrifugation steps (C and D) to obtain clarified plasma (E) handled by using an automatic platform (F) to extract cfDNA (G).

Acknowledgments

M. La Mantia and V. Gristina contributed to the current work under the Doctoral Program in Experimental Oncology and Surgery, University of Palermo.

References

[39] Malapelle U, Pisapia P, Rocco D, Smeraglio R, di Spirito M, Bellevicine C, et al. Next generation sequencing techniques in liquid biopsy: focus on non-small cell lung cancer patients. Transl Lung Cancer Res 2016;5:505−10.

[40] Rolfo C, Mack PC, Scagliotti GV, Baas P, Barlesi F, Bivona TG, et al. Liquid biopsy for advanced non-small cell lung cancer (NSCLC): a statement paper from the IASLC. J Thorac Oncol 2018;13:1248−68.

[41] Lam NY, Rainer TH, Chiu RW, Lo YM. EDTA is a better anticoagulant than heparin or citrate for delayed blood processing for plasma DNA analysis. Clin Chem 2004;50:256−7.

[42] Sherwood JL, Corcoran C, Brown H, Sharpe AD, Musilova M, Kohlmann A. Optimised pre-analytical methods improve KRAS mutation detection in circulating tumour DNA (ctDNA) from patients with non-small cell lung cancer (NSCLC). PLoS One 2016;11:e0150197.

[43] Malapelle U, Mayo de-Las-Casas C, Rocco D, Garzon M, Pisapia P, Jordana-Ariza N, et al. Development of a gene panel for next-generation sequencing of clinically relevant mutations in cell-free DNA from cancer patients. Br J Cancer 2017;116:802−10.

[44] Medina Diaz I, Nocon A, Mehnert DH, Fredebohm J, Diehl F, Holtrup F. Performance of streck cfDNA blood collection tubes for liquid biopsy testing. PLoS One 2016;11:e0166354.

[45] Schmidt B, Reinicke D, Reindl I, Bork I, Wollschläger B, Lambrecht N, et al. Liquid biopsy—performance of the PAXgene® blood ccfDNA tubes for the isolation and characterization of cell-free plasma DNA from tumor patients. Clin Chim Acta 2017;469:94−8.

[46] Toro PV, Erlanger B, Beaver JA, Cochran RL, VanDenBerg DA, Yakim E, et al. Comparison of cell stabilizing blood collection tubes for circulating plasma tumor DNA. Clin Biochem 2015;48:993−8.

[47] Sorber L, Zwaenepoel K, Jacobs J, De Winne K, Goethals S, Reclusa P, et al. Circulating cell-free DNA and RNA analysis as liquid biopsy: optimal centrifugation protocol. Cancers (Basel) 2019;11:458.

[48] Page K, Guttery DS, Zahra N, Primrose L, Elshaw SR, Pringle JH, et al. Influence of plasma processing on recovery and analysis of circulating nucleic acids. PLoS One 2013;8:e77963.

[49] Mauger F, Dulary C, Daviaud C, Deleuze JF, Tost J. Comprehensive evaluation of methods to isolate, quantify, and characterize circulating cell-free DNA from small volumes of plasma. Anal Bioanal Chem 2015;407:6873−8.

[50] Sorber L, Zwaenepoel K, Deschoolmeester V, Roeyen G, Lardon F, Rolfo C, et al. A comparison of cell-free DNA isolation kits: isolation and quantification of cell-free DNA in plasma. J Mol Diagn 2017;19:162−8.

Further reading

Froelich MF, Capoluongo E, Kovacs Z, Patton SJ, Lianidou ES, Haselmann V. The value proposition of integrative diagnostics for (early) detection of cancer. On behalf of the EFLM interdisciplinary Task and Finish Group "CNAPS/CTC for early detection of cancer" [published online ahead of print, 2022 Feb 24]. Clin Chem Lab Med 2022. Available from: https://doi.org/10.1515/cclm-2022-0129.

Rolfo C, Mack PC, Scagliotti GV, et al. Liquid biopsy for advanced non-small cell lung cancer (NSCLC): a statement paper from the IASLC. J Thorac Oncol 2018;13 (9):1248−68. Available from: https://doi.org/10.1016/j.jtho.2018.05.030.

Russo A, Incorvaia L, Capoluongo E, et al. The challenge of the molecular tumor board empowerment in clinical oncology practice: a position paper on behalf of the AIOM- SIAPEC/IAP-SIBioC-SIC-SIF-SIGU-SIRM Italian Scientific Societies. Crit Rev Oncol Hematol 2022;169:103567. Available from: https://doi.org/10.1016/j.critrevonc.2021.103567.

SUBCHAPTER 6.4

CTC and exosome: from the enrichment to the characterization

E. Lianidou[1],*, M. Castiglia[2],* and S. Taverna[3],*

[1]Analysis of Circulating Tumor Cells Laboratory, Department of Chemistry, University of Athens, Athens, Greece
[2]Department of Surgical, Oncological, and Oral Sciences, University of Palermo, Palermo, Italy
[3]Institute for Biomedical Research and Innovation (IRIB-CNR), National Research Council of Italy, Palermo, Italy

Learning objectives

By the end of the chapter the reader will:
- Know what liquid biopsy (LB) is and its potential application
- Have learned technical issues concerning exosomes and circulating tumor cells enrichment and characterization

6.4.1 Introduction

Cancer and its genetic profile change constantly acquiring new mutations and treatment resistance. The stochastic nature of clonal growth results in intra-tumoral heterogeneity and causes molecular sampling limitations in conventional tissue biopsy (TB). TB is an invasive procedure associated with several complications, particularly for patients on antiangiogenic treatment [51].

Tumor genetic characterization is important for therapy selection and prognosis and is based on TB of primary tumor or metastases. TB is a snapshot of cancer but cannot cover tumor heterogeneity. Since tumor masses are a heterogeneous agglomeration of cell subpopulations each carrying a distinct

* These authors equally contributed.

set of molecular alterations. Cancer biomarkers can reflect the molecular status and heterogeneity of tumors since these materials are released from different tumor sites or metastatic lesions. Moreover, different metastatic at diverse body sites can have an altered molecular profile [52].

Precision oncology is based on the idea that tumor biomarkers are predictive of disease phenotype, clinical outcomes, and therapy responses [53]. Precision oncology in cancer clinical management may be achieved with a new tool: LB [54].

LB is a real-time representative of tumors and is a minimally invasive biopsy method that has attracted scientists and oncologists over the past decades [55]. The evaluation of biomarkers in biological fluids is interesting, especially in cancer, because genetic information can be obtained relatively easily without the need for invasive surgery. The body fluids, such as blood, urine, cerebrospinal fluid, and saliva, contain biomarkers that mirror the patient-specific pathological information. In addition, LB allows a noninvasive sampling, brings forward new possibilities for cancer diagnosis and care, and opens the possibility to monitor the treatment response at therapies.

LB is useful for its application in different types of tumors such as lung, colorectal, prostate, melanoma, breast, and pancreatic cancer [56].

The principal components of LB are: circulating tumor DNA (ctDNA), extracellular vesicle (EVs) in particular exosomes (EXOs), ctRNA, circulating proteins, circulating tumor cells (CTCs), and tumor-educated platelets (Fig. 6.4.1) [57].

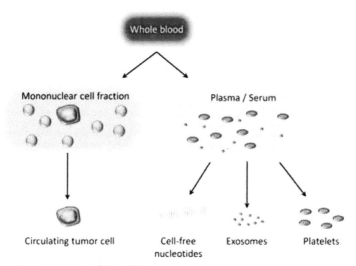

Figure 6.4.1 Component of liquid biopsy.

Qiu and colleagues suggested a novel analytical framework, called "integrated liquid biopsy," that could combine the conventional LB data with clinical requirements [57]. This framework is defined as integrating multiple LB biomarkers and detection approaches for enhanced analytical sensitivity and specificity to improve cancer management. A single biomarker can be insufficient in finding most cancers accurately. The integrated LB using multiple markers might be an encouraging method to help early detection and cancer treatment [58,59]. The introduction of NGS in the clinical oncology field has provided oncologists with a large amount of genomic information, which can be incorporated into clinical decision-making. In the era of precision oncology, cancer treatment is becoming more personalized.

A LB is also a new tool in the search for predictive and prognostic biomarkers in immunotherapy. ctDNA levels can have a prognostic and predictive value in patients with advanced tumors treated with immunotherapy [60]. In a LB, EXOs have an advantage compared with other sources such as CTCs and ctDNA due to the high stability and sufficient concentration in the circulation 28745363 [61]. The EVs contain a variety of contents, such as protein and miRNAs, which can be used as biomarkers for diseases. Moreover, EXOs contain high levels of miRNAs that contribute to immunoregulation and EXO-RNAs can be used as diagnostic biomarkers. Furthermore, EXOs derived from cancer cells and patients' blood are enriched in circular RNAs, which may serve as a new class of exosomal cancer biomarkers [62−65].

6.4.2 Exosome enrichment

Exosome isolation remains a remarkable technical challenge. The methods most commonly used are ultracentrifugation and commercial kits. The commercial methods consist of precipitation or size-based filtration techniques that are simple to use at low cost, and skilled personnel or facilities are not required. The limits of size exclusion techniques are the absence of specificity, pore-clogging, and shear stress-induced damage [66].

The current gold-standard procedure to isolate EXOs from blood remains the ultracentrifugation which is neither specific nor manageable. This procedure requires rather large sample volumes and is difficult to use and standardize, further involves preliminary steps such as various centrifugations at different speeds. The conventional procedure of ultracentrifugation is based on one step of centrifugation at 100.000 x g. This method has been improved by adding a second step of ultracentrifugation. The samples were collected and after centrifugation at different speeds to eliminate dead cells, cellular debris, and large vesicles, were ultra-centrifuged twice. This protocol, despite doubling the time required and loss of vesicles amount, allows obtaining EXO suspensions cleaner than EXO isolated by a single step of ultracentrifugation [67].

The aim of researchers, working in this field, is to find reliable methods with high specificity, able to distinguish different populations of vesicles; to avoid procedures time consuming, requiring skilled personnel, specific laboratory facilities, and particular benchtop equipment.

Recently, a growing number of papers report emerging technologies such as solid-phase separation including biosensors, and immune magnetic and microfluidic devices to avoid ultracentrifugation.

Immunoaffinity-based methods are potentially selective, highly sensitive, better protect the samples, and can be used to collect EXO used small volumes. These techniques can use EXOs or tumor-targeting antibodies bound on magnetic beads' surfaces or in microfluidic channels. The magnetic beads are usually coated with antibodies against exosomal markers, such as CD9 or CD63, allowing to isolate a specific vesicle population [68]. Immunoaffinity strategies might be also improved with a nanostructured polymer surface that increases binding avidity. It was reported that a dendrimer surface detects tumor EXOs at high sensitivity and specificity, indicating its possible use as a new LB platform [69]. Moreover, microfluidic devices using a specific immunoassay with magnetic microparticles were designed for the isolation of circulating EXOs; however, most of these devices are in an early development stage and not tested with a blood sample from a cancer patient [70]. Since LB is a good technique for early cancer diagnosis, the improvement of new methods to collect EXOs is ongoing. The detection and quantification of EXOs as cancer biomarkers can be also performed by ELISA, nanoparticle tracking analysis, flow cytometry, and transmission electron microscopy. Overall, the techniques developed so far require further improvements that allow standardizing the procedures to avoid variations that depend on the laboratory and/or investigators.

6.4.3 Exosomes characterization

The presence of nano-sized EVs has been known for over 70 years 20273687. Until the 1980s, when for the first time EVs biogenesis was described, EVs origin, characterization, and function remained vague [71,72]. In 1987, Johnstone termed "exosomes," EVs with small diameter-size originated from multivesicular endosomes matured to late endosomes and released from cells in extracellular space 3597417.

Nowadays, EVs classification is based on their intracellular origin, diameter size, and releasing mechanism. EVs are released constitutively by all cytotypes and have been collected by several biological fluids such as blood, malignant effusion, saliva, urine, breast milk, cerebrospinal, amniotic, and seminal fluid.

The two families of EVs, better described are EXOs and microvesicles that with their lipid bilayer protect the contents from degradation. Microvesicles have a diameter size of 150−1000 nm and are shed directly from cell membranes [73]. EXOs are small intraluminal vesicles (ILVs) with a diameter of 30−150 nm that originate from multivesicular bodies (MVBs) of the endocytic system and are released into extracellular space after the fusion of MVBs with the cell membrane.

EXOs are released with a regulated sorting mechanism, the endosomal sorting complex required for transport (ESCRT) was first described to regulate EXOs formation inside MVBs. ESCRT is composed of four complexes with other auxiliary proteins. ESCRT-0 controls the cargo clustering in ubiquitin-dependent-manner. ESCRT-I and II contribute to bud formation in MVBs and ESCRT-III induce vesicle shedding from MVB membranes. Moreover, Alix, a marker of EXOs is involved in their biogenesis; Alix binds to ESCRT-III and carries unubiquitinated cargoes to ILVs by directly binding to cargo or delivering syndecans and tetraspanin CD63 indirectly [74]. Alternative pathways for EXO formation were reported, for example, cholesterol induces vesicle secretion in Flotillin-2 dependent manner; phospholipase D2 increases ILVs production in MVB enhancing the inward curvature of the MVB membrane.

The molecules contained within EVs depend on the cell type and mirror origin and state of the parental cells. EXOs contain several proteins such as integrins, tetraspanin, transmembrane receptors, cytoskeletal (actin, myosin, tubulin) and heat shock proteins, MHC class I and II, cytokines, growth factors, and other bioactive molecules including lipids and nucleic acids, able to influence biological functions and phenotype of target cells. Furthermore, exosomal nucleic acids (mRNAs, miRNAs, noncoding RNAs long, circ-RNAs and double-strand DNA) have been widely described in different tumors as a shuttle of genetic information in the horizontal manner [75−78]. Experimental data indicates that exosomal nucleic acids have effects not only on cancer cells but create a bridge between cancer cells and the tumor microenvironment [79].

Jeppesen and colleagues propose a "reassessment" of exosome composition because by utilizing high-resolution density gradient fractionation and direct immunoaffinity capture to characterize RNA, DNA, and protein content in EXOs and other nonvesicle material, they showed that extracellular RNA, RNA-binding proteins, and other cellular proteins are differentially expressed in EVs and nonvesicle compartments. For example, they indicated that Argonaute 1−4, glycolytic enzymes such as GAPDH, and some cytoskeletal proteins were not detected in EXOs with these techniques [80].

EXOs can interact with target cells in different ways, by direct fusion of the membranes, binding to surface receptors, by releasing their contents into an extracellular milieu with a mechanism pH-dependent. EXOs uptake by target cells can be facilitated by micropinocytosis, clathrin-independent endocytosis, surfing on filopodia [81], or through filopodia extensions from cell membranes.

EVs have been described for the first time as "cellular garbage bins" to eliminate obsolete molecules from cells. Nowadays, a growing number of studies reported that EXOs have important functional roles in intercellular communication and signaling and are involved in a plethora of biological processes [82,83]. EXOs contribute to pathological remodeling [84] of organs and participate in tumor angiogenesis regulation [85,86]. EXOs internalization by endothelial cells in response to enhanced EXOs release caused by intra-tumoral hypoxia stimulates angiogenesis and impairs the structural integrity of endothelium [87,88]. These vesicles contain active proteases capable of extracellular matrix (ECM) degradation, contributing to ECM remodeling. EXOs are also considered protumorigenic and have a key role in cancer progression [89] inducing tumor growth and premetastatic niche formation [90−92]. It was also reported that EXOs are the main actors in metastasis induction at distant sites [93−95].

A very interesting and attractive research field concerns the involvement of EXOs in immune system regulation.

6.4.3.1 Role of exosomes in immunotherapy

EXO-effects in tumor immunology are dynamic and intricate extending from tumor antigen presentation control to polarization of tumor immunity [96]. The advantage of EXOs has the high stability in circulation and the intrinsic ability of horizontal cargo transfer and low toxicity and immunogenicity. CD47, a widely expressed integrin-associated transmembrane protein on EXOs, can avoid phagocytosis by circulating monocytes, thus promoting the delivery of their cargos. CD47 interacts with its ligand signal regulatory protein α (SIRPα or CD172a) on macrophages, which induces a signal of phagocytosis inhibition. The EXOs contain a plasma membrane and membrane binding proteins, which could contribute to their diminished clearance from the circulation. Furthermore, EXOs can cross the blood-brain barrier they are also used as a novel strategy for tumor brain treatment [97].

EXOs have emerged as mediators between the immune system and cancer cell communication, with anti and protumorigenic roles that reflect the functional heterogeneity of EXOs in the tumor microenvironment [86]. EXOs released by cancer cells can activate NK cells via the presentation of stress protein HSP70 in a direct way [98]. Furthermore, EXOs released by mast cells can activate T and B cells via DC differentiation, indirectly [99]. These data indicate that EXO-immune activities support their role in promoting antitumor immune responses. Conversely, EXOs have also a role in immune escape by impairing DC maturation through IL-6 increased expression in BM dendritic precursor cells [100]. Cancer-EXOs can also inhibit NK cell proliferation and cytotoxic functions downregulating NK group 2, member D, and NKG2D [101]. EXOs released from dendritic cells (DCs) can activate T and B cells and may serve as a source of tumor antigens that can be presented to activated T cells.

Cancer-EXOs can alter T cell biology causing T cell apoptosis;[102] it was reported that Fas ligand on EXOs induces apoptosis of Fas + T cells [103]. EXOs may suppress T cell receptor activity and regulate the transcriptome of T cells regulatory and effector (Tregs and Teffs) [104]. TGFβ1 in cancer EXOs is reported to induce Tregs [105−107]. Overall, these data indicate the role of EXOs in antitumor immune responses. The treatment of lung cancer patients with dexosomes, DC-derived EXOs, can activate the antitumor immune response.

Moreover, EXOs are regulatory elements through which cancer cells can communicate and reprogram the immune cells population in the tumor microenvironment [108,109]. Cancer-EXOs carry bioactive PD-L1 on their surface and can suppress the immune response [105,110].

In normal physiological conditions, the activation of the PD-1/PD-L1 signaling pathway is involved in the induction and maintenance of peripheral tolerance, preservation of T cell immune homeostasis, and avoiding hyperactivation of the immune system [111]. In cancer, PD-L1 interacts with its receptor, PD-1, transmitting a negative signal to control T cell-mediated immune response. PD-1/PD-L1 pathway can block the T cell cycle at the G1 phase rather than directly causing apoptosis [112]. PD-1/PD-L1 pathway could also induce inhibition of T cell responses by inducible Tregs [113].

It was reported that EXOs collected from plasma of patients with head and neck squamous cell carcinomas (HNSCC) have immunosuppressive properties and play a role in tumor progression. EXOs carry PD-L1 and PD-1, but it was not clear if PD-1 and PD-L1 contained in EXOs were biologically active and responsible for the immune-inhibitory effects [114].

Moreover, EXOs of patients with HNSCC that shuttle PD-L1 can induce T-cell dysfunction upon coincubation of these EXOs with activated CD8 + T cells. Although soluble PD-L1 levels were elevated in plasma patients, only exosome-bound PD-L1 levels correlated with disease activity and clinic conditions of the patients. Anti-PD-1 antibodies block the suppression induced by PD-L1 + EXOs in activated T cells. Exosomal PD-L1 + emerges as a potential marker of immune dysfunction and disease progression [115]. Since cancer cells release large amounts of EXOs, compared with normal cells, they are a good source of biomarkers.

EXOs released by metastatic melanoma cells stimulated by interferon-γ expressed more PD-L1 on these vesicles and inhibited antitumor responses [116]. Melanoma cells release PD-L1-positive EVs into the tumor microenvironment and circulation; since exosomal PD-L1-mediated T cell inhibition can be blocked by both anti-PD-L1 and PD-1 antibodies. Circulating exosomal PD-L1 is considered an effective predictor for anti-PD-1 therapy in melanomas [116].

The features of EXOs such as stability and ability to stimulate the immune responses against cancer cells and target specific cells in distant tissues make these vesicles a suitable vehicle for cancer chemotherapy and immunotherapy compounds [117,118]. It was demonstrated that EXOs participate in the establishment of an immunosuppressive environment found associated with pancreatic ductal adenocarcinoma (PDAC) [119]. One of the therapeutic approaches frequently used for the treatment of PDAC patients combines the use of chemotherapy with inhibitors of immunosuppression and agents that stimulate immune cytotoxic responses against cancer cells [120,121]. The use of EXOs in PDAC immunotherapy can improve the survival and life quality of PDAC patients. Nowadays, two reports used exosome-based delivery of immunomodulatory agents to tumors. a preclinical study suggests the administration of vaccines containing DCs that enclosed tumor-derived EXOs to induce T cell-driven antitumor immune responses [122]. The other study showed *in vitro* that EXOs derived from Panc-1 cells expressing miR-155 and miR-125b2 could reprogram M2 macrophages back to an M1 phenotype, which could then impair tumor growth [123].

Although EXOs are interesting biological entities that could offer oncologists important new therapeutic modalities for cancer treatment, several challenges need to be overcome before their full clinical use, and further studies on the role of EXOs in cancer and immune modulation need to be carried out to use this knowledge for new exosome-based therapies development.

In the era of precision oncology in the clinical management of cancer, LB, due to its powerful real-time approach, represents an important innovation for the molecular monitoring of this dynamic tumor disease. The abundant number of EXOs that are released by tumor cells compared to nontumorigenic cells makes these vesicles very promising for the early detection of cancer and its monitoring to evaluate the efficacy of anticancer therapy.

6.4.4 Circulating tumor cells enrichment methods

Most of the existing technologies for CTCs isolation are based on a two-step process that starts with cell enrichment followed by detection. CTCs capturing can be obtained by exploiting CTCs physical properties (charge, size, elasticity, and density) [124−129] as well as various biological features (cellular functions [130] and tumor-specific surface proteins [131−136].

Dielectrophoresis (DEP) is a well-known electrokinetic phenomenon that arises when a neutral object is exposed to a non-uniform electric field [137,138]. To improve DEP discrimination capacity, the dielectrophoretic field-flow fractionation (DEP-FFF) was developed [139]. In the field of medicine and biology, the DEP-FFF has been demonstrated in many applications, including the isolation of CTCs in cancer patients [124]. Cells exposed to DEP forces are positioned in a hydrodynamic flow profile according to their dielectric properties and they are carried through the channel at characteristic heights and speeds, with dissimilar particle types emerging separated in time and/or space [125]. The DEP-FFF is based on the use of a chamber that contains a flow channel having a floor lined with interdigitated 50-μ wide gold-on-copper electrodes [124].

Cells are injected into the chamber by syringe, thereafter an electrical signal is applied to the electrode. The combination of dielectrophoretic, sedimentation, and hydrodynamic forces influence the cell position in the hydrodynamic flow profile; as a result, cells having different properties are eluted at different times from the chamber [124]. Therefore, according to the specific dielectric properties, CTCs are separated from blood cells. The DEP-FFF can process approximately 30 million cells within 30 min, therefore, showing both high throughput and recovery rates [140]. CTCs isolated with DEP-FFF can be cultured and used for several downstream analyses, however, it requires very specific parameters such as cell type and electric field frequency.

Size-based technologies take advantage of the greater CTCs dimension compared to blood cells, thus by using a specific separation device it is possible to easily isolate CTCs. However, one main drawback of this approach is that it fails to isolate smaller CTCs. Among the most known and used size-based technologies we can include the ISET device [128], MetaCell filtration device [141], and Dean Flow Fractionation (DFF) [142], Parsotrix method [143], and ScreenCell devices [144].

The ISET is carried out using a filtration module [145] on a membrane which is composed of 8μm-diameter cylindrical pores, blood sample is first diluted using a specific buffer and then filtered by gentle aspiration using a vacuum pump. The membrane which contains the isolated cells is then washed and can be used for further characterization (e.g., staining with hematoxylin and eosin or May-Grumwald-Giemsa).

In the MetaCell method, the enrichment process is based on the filtration, driven by capillarity, of peripheral blood through a porous polycarbonate membrane (8 μm diameter). Therefore, at the end of filtration which speed depends on the natural blood viscosity, cells are viable and have not been damaged by any fixatives or lysing solutions. Consequently, the membrane can be directly transferred on culture plates to proceed with further investigations (e.g., tests assessing chemosensitivity) [146].

The DFF is a high throughput system that uses a spiral microchannel with inherent centrifugal forces for continuous CTCs separation. Isolated CTCs can be used for various downstream applications such as CTCs enumeration and culturing [142].

The Parsotrix Cell separation system exploits size and CTCs deformability properties to separate them from blood cells and is based on the use of specific microfluidic cassettes that guarantee reproducibility, high capture efficiency, and produce highly enriched, viable CTCs that are amenable to a multitude of downstream analyses, including the isolation and interrogation of single cells [143].

The ScreenCell devices use filters to quickly isolate and sort tumor cells by size; in detail, the ScreenCell Cyto device allows the isolation of rare, fixed, tumor cells that are morphologically well preserved and can be used for immunocytochemistry and FISH assays. The other device, the ScreenCell CC, is instead designed to isolate viable CTCs that be cultured. Finally, the ScreenCell MB device filter is meant to be used to obtain high-quality genetic materials from CTCs [144].

Besides the differences in terms of size between CTCs and normal cells, CTCs are known to express specific protein markers on their surface that are not found in other circulating cells. This specific biological feature can therefore be used to isolate them from a blood sample. It is known that CTC selectively expresses epithelial cell adhesion molecules (EpCAM) and, indeed, this is one of the main protein markers used for CTC positive selection. Moreover, to reduce the possibility of contamination with another circulating cell type, the positive selection can be coupled with a negative selection by discarding cells expressing the CD45 marker. In the immunebead-based assays, magnetic beads are coated with specific antibodies against the above-mentioned markers, and cells are subsequently removed by applying a magnetic field.

The first FDA-approved method for CTCs enrichment and detection in breast, prostate, and colorectal cancer is the immunobead-based assay CellSearch system [147,148]. With this platform, $CD45^+$-leucocytes are negatively selected and excluded from the analysis whereas EpCAM antibodies are used to select CTCs whose nuclei are also evaluated using DAPI stains. The CellSearch system is considered the gold standard and is therefore commonly used as a benchmark for other detection and enrichment methods. The common drawback of EpCAM targeting assay is the impossibility to detect CTCs that undergo EMT since they are EpCAM negative.

Other popular immunebead-based assays include the AdnaTest, MagSweeper, and IsoFlux. The AdnaTest is a two-step process that enables both immunomagnetic cell selection and detection of cancer-specific mRNA markers for prostate, breast, and lung cancer [149]. Cell selection is obtained using a combination of tumor-associated markers including EpCAM, CA15−3 (Muc1), and the presence of Her2 receptors for the breast cancer test and PSMA, PSA, and EGFR for prostate cancer test. It is known that EMT commonly occurs in lung cancer, to overcome this problem the lung cancer AdnaTest provides, in addition to anti-EpCAM, antibodies against Her2 and EGFR, resulting in a strong increase in performance. The MasSweeper uses a robotically-controlled magnetic rod able to isolate CTCs by sweeping through wells containing samples premixed with antibody-coated magnetic beads [136,150]. The IsoFlux system is a microfluidic device for CTCs immunomagnetic isolation [151]. Briefly, a blood sample is first labeled with immunomagnetic beads containing antiEpCAM antibodies, then injected into the sample well and conveyed in a microfluidic channel reaching the isolation zone. In this zone, magnetically labeled cells are captured and finally collected in a dedicated recovery tube.

Based on the knowledge acquired regarding CTC surface markers, in recent years several microdevices have been developed. These microdevices have the advantage to be easily introduced into clinical practice. Among the microdevices that work ex vivo, we can include the CTC iChip [152,153], the Cluster Chip [154], and the Herringbone CTC-Chip [135]. All these devices work with various antibody "cocktails" in addition to standard anti-EpCAM proteins, enabling more precise isolation of CTCs. Moreover, some of them can combine immunobead-based and size-based methods such as the CTC iChip and the CTC-chip. A very interesting device, which can be applied in vivo is the GILUPI GmbH CellCollector [155]. It is composed of a stainless-steel 16 cm wire coated with anti-EpCAM antibodies that can be directly placed in a vein enabling the processing of approximately 1.5 L of blood in 30 min.

In summary, size-based CTC isolation methods allow a high throughput but with limited clinical applicability due to the high heterogeneity of recovered CTCs. Nevertheless, one of the main advantages of using these methods is the possibility to capture cells that have undergone EMT. On the other hand, immunobead-based assays allow isolation of CTC that expresses specific surface markers (positive selection) and at the same time, they assure the removal of all circulating cells that expresses CD45 marker (negative selection). Nevertheless, also these methods have several limitations, such as the inability to select atypical cells, which do not express the selected markers, resulting in CTC numbers underestimation but also the impossibility to detect a CTC of mesenchymal origin.

Following CTC enrichment using one of the aforementioned methods, it is possible to proceed with several downstream including fluorescence in situ hybridization (FISH), immunocytochemistry (ICC), RT-PCR/qRT-PCR for quantifying specific RNA, and Next Generation Sequencing (NGS) for whole-genome analysis. The ICC is the most used method due to the very ease of use as well as the reduced cost of the analysis. Downstream application targeting mRNA analysis offers a highly specific detection method but can lead to false-negative and false-positive, especially when working with not highly purified samples. The same drawback is faced for FISH analysis, indeed for example cells isolated via CellSearch appear to be more apoptotic and less likely to generate FISH signals. FISH analyses maybe, instead, be feasible after using a filtration enrichment technique [156]. Finally, by using NGS technologies it is possible to molecular profiling of a single CTC. Nevertheless, this approach requires highly pure samples and due to its complexity, high costs of the analysis, and the lack of technique standardization, the method is not easily applicable in clinical practice whereas it earns enormous potential for translational research.

6.4.4.1 Role of circulating tumor cells in immunotherapy

Immunotherapy in cancer treatment is a rapidly evolving field, and it has already revolutionized the treatment strategies in many solid tumors. The only validated predictive biomarker for first-line immunotherapy is the evaluation of PD-L1 protein expression in tumor tissue assessed by immunohistochemistry (IHC). Nevertheless, despite the enormous efforts that have done throughout molecular pathology laboratories to standardize PD-L1 IHC, there are still several technical and biological limitations in using this biomarker, and its assessment can turn out to be quite challenging for the pathologist [157,158].

One of the main drawbacks of PD-L1 evaluation is its heterogeneous intra- and intertumor expression [159], meaning that PD-L1 status can be underestimated in small biopsies that may not be representative of the entire tumor. From this and other reasons arises the need to investigate new biomarkers and to develop new noninvasive tools that allow strict patient monitoring during treatment trying to overcome issues related to heterogeneity. LB encompasses this need. LB has been already implemented in clinical practice, mainly in thoracic oncology for targeted therapy. The next step is therefore to increase the effort to bring LB also in immunotherapy.

CTCs directly originated from more than one tumor lesion and therefore they can provide a more comprehensive representation of PD-L1 expression than tumor tissue. It has been widely demonstrated that PD-L1 status can be evaluated in CTCs in many different tumor types. Iliè M. et al. reported that in lung cancer PD-L1 expression in CTCs and white blood cells is correlated with the one reported in tumor tissue [160]. This result clearly and strongly supports the potential use of CTCs as a noninvasive real-time biopsy to evaluate PD-L1 expression. PD-L1 evaluation in CTCs can provide prognostic information in patients treated with anti-PD1 nivolumab, as demonstrated by Guibert et al. [161]. They showed that finding a high number of PD-L1$^+$CTCs in pretreatment samples is associated with a bad prognosis in patients treated with nivolumab [161]. Moreover, serial assessment of PD-L1 expressing CTCs can be useful for treatment response monitoring. Nicolazzo et al. showed that the persistence of PD-L1$^+$CTCs after 6 months of treatment with nivolumab is associated with progressive disease and treatment failure; conversely, all patients with no PD-L1 expressing CTCs after 6 months have all obtained clinical benefit [162]. Similarly, in colorectal and prostate cancer patients undergoing treatment, the detection of CTCs expressing nuclear PD-L1 is significantly associated with short survival durations [163].

Despite these promising results, prospective clinical trials that assess the use of PD-L1 expressing CTCs for the initial treatment decision as well as during monitoring to predict response are urgently needed. CTCs enrichment and detection remains the main technical challenge for CTCs to improve in the domain of immunotherapy. Nevertheless, it is reasonable to suggest that the incorporation of CTCs analysis in a clinic can improve the standard cancer staging criteria as well as patients' monitoring during different disease phases (diagnosis, treatment response, treatment resistance, disease progression).

6.4.5 Key points

- Biomarkers allow clinicians to switch from standardized medicine to a more tailored approach;
- Agnostic biomarkers allow to select of patients that can benefit from specific drugs, independently from tumor site or histology;
- Modern oncology is perfectly complemented by molecular biology

References

[51] Adams DL, et al. Sequential tracking of PD-L1 expression and RAD50 induction in circulating tumor and stromal cells of lung cancer patients undergoing radiotherapy. Clin Cancer Res 2017;23(19):5948−58.
[52] Dagogo-Jack I, Shaw AT. Tumour heterogeneity and resistance to cancer therapies. Nat Rev Clin Oncol 2018;15(2):81−94.
[53] Li X, Warner JL. A review of precision oncology knowledgebases for determining the clinical actionability of genetic variants. Front Cell Dev Biol 2020;8:48.

[54] Doroshow DB, Doroshow JH. Genomics and the history of precision oncology. Surg Oncol Clin N Am 2020;29(1):35−49.

[55] Geeurickx E, Hendrix A. Targets, pitfalls and reference materials for liquid biopsy tests in cancer diagnostics. Mol Asp Med 2020;72:100828.

[56] Fernández-Lázaro D, et al. Liquid biopsy as novel tool in precision medicine: origins, properties, identification and clinical perspective of cancer's biomarkers. Diagnostics (Basel) 2020;10(4).

[57] Qiu J, et al. Refining cancer management using integrated liquid biopsy. Theranostics 2020;10(5):2374−84.

[58] van Galen P, et al. Single-Cell RNA-Seq reveals AML hierarchies relevant to disease progression and immunity. Cell 2019;176(6):1265−81 e24.

[59] Yokoi A, et al. Integrated extracellular microRNA profiling for ovarian cancer screening. Nat Commun 2018;9(1):4319.

[60] Remon J, et al. Liquid biopsy in oncology: a consensus statement of the Spanish Society of Pathology and the Spanish Society of Medical Oncology. Clin Transl Oncol 2020;22(6):823−34.

[61] Im H, et al. Novel nanosensing technologies for exosome detection and profiling. Lab Chip 2017;17(17):2892−8.

[62] Zhu N, et al. Endothelial-specific intron-derived miR-126 is down-regulated in human breast cancer and targets both VEGFA and PIK3R2. Mol Cell Biochem 2011;351(1−2):157−64.

[63] Fanale D, et al. Circular RNA in exosomes. Adv Exp Med Biol 2018;1087:109−17.

[64] Reclusa P, et al. Exosomes as diagnostic and predictive biomarkers in lung cancer. J Thorac Dis 2017;9(Suppl 13):S1373−82.

[65] Pucci M, et al. Extracellular vesicles as miRNA Nano-shuttles: dual role in tumor progression. Target Oncol 2018;13(2):175−87.

[66] Boukouris S, Mathivanan S. Exosomes in bodily fluids are a highly stable resource of disease biomarkers. Proteom Clin Appl 2015;9(3−4):358−67.

[67] Reclusa P, et al. Improving extracellular vesicles visualization: from static to motion. Sci Rep 2020;10(1):6494.

[68] Jia Y, et al. Exosome: emerging biomarker in breast cancer. Oncotarget 2017;8 (25):41717−33.

[69] Poellmann MJ, et al. Immunoavidity-based capture of tumor exosomes using poly (amidoamine) dendrimer surfaces. Nano Lett 2020;20(8):5686−92.

[70] Sierra J, et al. Sensor-integrated microfluidic approaches for liquid biopsies applications in early detection of cancer. Sensors (Basel) 2020;20(5).

[71] Pan BT, Johnstone RM. Fate of the transferrin receptor during maturation of sheep reticulocytes in vitro: selective externalization of the receptor. Cell 1983;33(3):967−78.

[72] Harding C, Heuser J, Stahl P. Receptor-mediated endocytosis of transferrin and recycling of the transferrin receptor in rat reticulocytes. J Cell Biol 1983;97(2):329−39.

[73] Colombo M, Raposo G, Théry C. Biogenesis, secretion, and intercellular interactions of exosomes and other extracellular vesicles. Annu Rev Cell Dev Biol 2014;30:255−89.

[74] Mashouri L, et al. Exosomes: composition, biogenesis, and mechanisms in cancer metastasis and drug resistance. Mol Cancer 2019;18(1):75.

[75] Rolfo C, et al. Liquid biopsies in lung cancer: the new ambrosia of researchers. Biochim Biophys Acta 2014;1846(2):539−46.

[76] Taverna S, et al. Exosomal shuttling of miR-126 in endothelial cells modulates adhesive and migratory abilities of chronic myelogenous leukemia cells. Mol Cancer 2014;13:169.

[77] Galvano A, et al. Detection of RAS mutations in circulating tumor DNA: a new weapon in an old war against colorectal cancer. A systematic review of literature and *meta*-analysis. Ther Adv Med Oncol 2019;11. p. 1758835919874653.

[78] Valadi H, et al. Exosome-mediated transfer of mRNAs and microRNAs is a novel mechanism of genetic exchange between cells. Nat Cell Biol 2007;9(6):654—9.

[79] Guo W, et al. Exosomes: new players in cancer (Review). Oncol Rep 2017;38 (2):665—75.

[80] Jeppesen DK, et al. Reassessment of exosome composition. Cell 2019;177 (2):428—45 e18.

[81] Heusermann W, et al. Exosomes surf on filopodia to enter cells at endocytic hot spots, traffic within endosomes, and are targeted to the ER. J Cell Biol 2016;213(2):173—84.

[82] Hessvik NP, Llorente A. Current knowledge on exosome biogenesis and release. Cell Mol Life Sci 2018;75(2):193—208.

[83] Giallombardo M, et al. Exosome-mediated drug resistance in cancer: the near future is here. Ther Adv Med Oncol 2016;8(5):320—2.

[84] Das A, et al. Exosomes as a storehouse of tissue remodeling proteases and mediators of cancer progression. Cancer Metastasis Rev 2019;38(3):455—68.

[85] Taverna S, et al. Role of exosomes released by chronic myelogenous leukemia cells in angiogenesis. Int J Cancer 2012;130(9):2033—43.

[86] Kalluri R. The biology and function of exosomes in cancer. J Clin Invest 2016;126 (4):1208—15.

[87] Zhou W, et al. Cancer-secreted miR-105 destroys vascular endothelial barriers to promote metastasis. Cancer Cell 2014;25(4):501—15.

[88] Taverna S, et al. Curcumin modulates chronic myelogenous leukemia exosomes composition and affects angiogenic phenotype via exosomal miR-21. Oncotarget 2016;7(21):30420—39.

[89] Osaki M, Okada F. Exosomes and their role in cancer progression. Yonago Acta Med 2019;62(2):182—90.

[90] Feng W, et al. Exosomes promote pre-metastatic niche formation in ovarian cancer. Mol Cancer 2019;18(1):124.

[91] Lu Z, et al. Epigenetic therapy inhibits metastases by disrupting premetastatic niches. Nature 2020;579(7798):284—90.

[92] Adem B, Vieira PF, Melo SA. Decoding the biology of exosomes in metastasis. Trends Cancer 2020;6(1):20—30.

[93] Taverna S, et al. Breast cancer derived extracellular vesicles in bone metastasis induction and their clinical implications as biomarkers. Int J Mol Sci 2020;21(10).

[94] Taverna S, et al. Amphiregulin contained in NSCLC-exosomes induces osteoclast differentiation through the activation of EGFR pathway. Sci Rep 2017;7(1):3170.

[95] Gener Lahav T, et al. Melanoma-derived extracellular vesicles instigate proinflammatory signaling in the metastatic microenvironment. Int J Cancer 2019;145(9):2521—34.

[96] Greening DW, et al. Exosomes and their roles in immune regulation and cancer. Semin Cell Dev Biol 2015;40:72—81.

[97] Xie F, et al. Extracellular vesicles in cancer immune microenvironment and cancer immunotherapy. Adv Sci (Weinh) 2019;6(24):1901779.

[98] Lancaster GI, Febbraio MA. Exosome-dependent trafficking of HSP70: a novel secretory pathway for cellular stress proteins. J Biol Chem 2005;280(24):23349—55.

[99] Skokos D, et al. Mast cell-derived exosomes induce phenotypic and functional maturation of dendritic cells and elicit specific immune responses in vivo. J Immunol 2003;170(6):3037—45.

[100] Yu S, et al. Tumor exosomes inhibit differentiation of bone marrow dendritic cells. J Immunol 2007;178(11):6867—75.

[101] Clayton A, et al. Human tumor-derived exosomes down-modulate NKG2D expression. J Immunol 2008;180(11):7249—58.

[102] Zöller M, et al. Immunoregulatory effects of myeloid-derived suppressor cell exosomes in mouse model of autoimmune alopecia areata. Front Immunol 2018;9:1279.

[103] Abusamra AJ, et al. Tumor exosomes expressing Fas ligand mediate CD8 + T-cell apoptosis. Blood Cell Mol Dis 2005;35(2):169−73.

[104] Salimu J, et al. Dominant immunosuppression of dendritic cell function by prostate-cancer-derived exosomes. J Extracell Vesicles 2017;6(1):1368823.

[105] Seo N, Akiyoshi K, Shiku H. Exosome-mediated regulation of tumor immunology. Cancer Sci 2018;109(10):2998−3004.

[106] Clayton A, et al. Cancer exosomes express CD39 and CD73, which suppress T cells through adenosine production. J Immunol 2011;187(2):676−83.

[107] Whiteside TL. Immune modulation of T-cell and NK (natural killer) cell activities by TEXs (tumour-derived exosomes). Biochem Soc Trans 2013;41(1):245−51.

[108] Becker A, et al. Extracellular vesicles in cancer: cell-to-cell mediators of metastasis. Cancer Cell 2016;30(6):836−48.

[109] Passiglia F, et al. Primary and metastatic brain cancer genomics and emerging biomarkers for immunomodulatory cancer treatment. Semin Cancer Biol 2018;52(Pt 2):259−68.

[110] Xie F, et al. The role of exosomal PD-L1 in tumor progression and immunotherapy. Mol Cancer 2019;18(1):146.

[111] Bardhan K, Anagnostou T, Boussiotis VA. The PD1:PD-L1/2 pathway from discovery to clinical implementation. Front Immunol 2016;7:550.

[112] Patsoukis N, et al. Selective effects of PD-1 on Akt and Ras pathways regulate molecular components of the cell cycle and inhibit T cell proliferation. Sci Signal 2012;5(230):ra46.

[113] Zhang SA, et al. Effect of EBI3 on radiation-induced immunosuppression of cervical cancer HeLa cells by regulating Treg cells through PD-1/PD-L1 pathway. Tumour Biol 2017;39(3). p. 1010428317692237.

[114] Ludwig S, et al. Suppression of lymphocyte functions by plasma exosomes correlates with disease activity in patients with head and neck cancer. Clin Cancer Res 2017;23(16):4843−54.

[115] Theodoraki MN, et al. Clinical significance of PD-L1. Clin Cancer Res 2018;24 (4):896−905.

[116] Chen G, et al. Exosomal PD-L1 contributes to immunosuppression and is associated with anti-PD-1 response. Nature 2018;560(7718):382−6.

[117] Kamerkar S, et al. Exosomes facilitate therapeutic targeting of oncogenic KRAS in pancreatic cancer. Nature 2017;546(7659):498−503.

[118] Batista IA, Melo SA. Exosomes and the future of immunotherapy in pancreatic cancer. Int J Mol Sci 2019;20(3).

[119] Basso D, et al. PDAC-derived exosomes enrich the microenvironment in MDSCs in a. Oncotarget 2017;8(49):84928−44.

[120] Sahin IH, et al. Immunotherapy in pancreatic ductal adenocarcinoma: an emerging entity? Ann Oncol 2017;28(12):2950−61.

[121] Guo S, et al. Immunotherapy in pancreatic cancer: unleash its potential through novel combinations. World J Clin Oncol 2017;8(3):230−40.

[122] Yu Z, et al. Pancreatic cancer-derived exosomes promote tumor metastasis and liver pre-metastatic niche formation. Oncotarget 2017;8(38):63461−83.

[123] Su MJ, Aldawsari H, Amiji M. Pancreatic cancer cell exosome-mediated macrophage reprogramming and the role of MicroRNAs 155 and 125b2 transfection using nanoparticle delivery systems. Sci Rep 2016;6:30110.

[124] Gascoyne PR, et al. Isolation of rare cells from cell mixtures by dielectrophoresis. Electrophoresis 2009;30(8):1388−98.

[125] Gascoyne PR. Dielectrophoretic-field flow fractionation analysis of dielectric, density, and deformability characteristics of cells and particles. Anal Chem 2009;81 (21):8878−85.

[126] Moon HS, et al. Continuous separation of breast cancer cells from blood samples using multi-orifice flow fractionation (MOFF) and dielectrophoresis (DEP). Lab Chip 2011;11(6):1118−25.

[127] Müller V, et al. Circulating tumor cells in breast cancer: correlation to bone marrow micrometastases, heterogeneous response to systemic therapy and low proliferative activity. Clin Cancer Res 2005;11(10):3678−85.

[128] Vona G, et al. Isolation by size of epithelial tumor cells: a new method for the immunomorphological and molecular characterization of circulatingtumor cells. Am J Pathol 2000;156(1):57−63.

[129] Zheng S, et al. 3D microfilter device for viable circulating tumor cell (CTC) enrichment from blood. Biomed Microdevices 2011;13(1):203−13.

[130] Alix-Panabières C. EPISPOT assay: detection of viable DTCs/CTCs in solid tumor patients. Recent Results Cancer Res 2012;195:69−76.

[131] Allard WJ, et al. Tumor cells circulate in the peripheral blood of all major carcinomas but not in healthy subjects or patients with nonmalignant diseases. Clin Cancer Res 2004;10(20):6897−904.

[132] Helzer KT, et al. Circulating tumor cells are transcriptionally similar to the primary tumor in a murine prostate model. Cancer Res 2009;69(19):7860−6.

[133] Lu YT, et al. NanoVelcro Chip for CTC enumeration in prostate cancer patients. Methods 2013;64(2):144−52.

[134] Riethdorf S, et al. Detection of circulating tumor cells in peripheral blood of patients with metastatic breast cancer: a validation study of the CellSearch system. Clin Cancer Res 2007;13(3):920−8.

[135] Stott SL, et al. Isolation of circulating tumor cells using a microvortex-generating herringbone-chip. Proc Natl Acad Sci U S A 2010;107(43):18392−7.

[136] Talasaz AH, et al. Isolating highly enriched populations of circulating epithelial cells and other rare cells from blood using a magnetic sweeper device. Proc Natl Acad Sci U S A 2009;106(10):3970−5.

[137] Chen CS, Pohl HA. Biological dielectrophoresis: the behavior of lone cells in a nonuniform electric field. Ann N Y Acad Sci 1974;238:176−85.

[138] Pohl HA, Crane JS. Dielectrophoresis of cells. Biophys J 1971;11(9):711−27.

[139] Huang Y, et al. Introducing dielectrophoresis as a new force field for field-flow fractionation. Biophys J 1997;73(2):1118−29.

[140] Sharma S, et al. Circulating tumor cell isolation, culture, and downstream molecular analysis. Biotechnol Adv 2018;36(4):1063−78.

[141] Bobek V, Kolostova K. Isolation and characterization of CTCs from patients with cancer of a urothelial origin. Methods Mol Biol 2018;1655:275−86.

[142] Hou HW, et al. Isolation and retrieval of circulating tumor cells using centrifugal forces. Sci Rep 2013;3:1259.

[143] Miller MC, et al. The Parsortix™ cell separation system-a versatile liquid biopsy platform. Cytometry A 2018;93(12):1234−9.

[144] Desitter I, et al. A new device for rapid isolation by size and characterization of rare circulating tumor cells. Anticancer Res 2011;31(2):427−41.

[145] Rostagno P, et al. Detection of rare circulating breast cancer cells by filtration cytometry and identification by DNA content: sensitivity in an experimental model. Anticancer Res 1997;17(4A):2481−5.

[146] Bobek V, et al. Circulating tumor cells in pancreatic cancer patients: enrichment and cultivation. World J Gastroenterol 2014;20(45):17163−70.

[147] Harouaka RA, et al. Flexible micro spring array device for high-throughput enrichment of viable circulating tumor cells. Clin Chem 2014;60(2):323−33.

[148] Kim MS, et al. SSA-MOA: a novel CTC isolation platform using selective size amplification (SSA) and a multi-obstacle architecture (MOA) filter. Lab Chip 2012;12(16):2874−80.

[149] Zieglschmid V, et al. Combination of immunomagnetic enrichment with multiplex RT-PCR analysis for the detection of disseminated tumor cells. Anticancer Res 2005;25(3A):1803−10.

[150] Ferreira MM, Ramani VC, Jeffrey SS. Circulating tumor cell technologies. Mol Oncol 2016;10(3):374−94.

[151] Harb W, et al. Mutational analysis of circulating tumor cells using a novel micro-fluidic collection device and qPCR assay. Transl Oncol 2013;6(5):528−38.

[152] Karabacak NM, et al. Microfluidic, marker-free isolation of circulating tumor cells from blood samples. Nat Protoc 2014;9(3):694−710.

[153] Ozkumur E, et al. Inertial focusing for tumor antigen-dependent and -independent sorting of rare circulating tumor cells. Sci Transl Med 2013;5(179):179ra47.

[154] Sarioglu AF, et al. A microfluidic device for label-free, physical capture of circulat-ing tumor cell clusters. Nat Methods 2015;12(7):685−91.

[155] Saucedo-Zeni N, et al. A novel method for the in vivo isolation of circulating tumor cells from peripheral blood of cancer patients using a functionalized and structured medical wire. Int J Oncol 2012;41(4):1241−50.

[156] Pailler E, et al. Detection of circulating tumor cells harboring a unique ALK rearrange-ment in ALK-positive non-small-cell lung cancer. J Clin Oncol 2013;31(18):2273−81.

[157] Tsao MS, et al. PD-L1 immunohistochemistry comparability study in real-life clinical samples: results of blueprint phase 2 project. J Thorac Oncol 2018;13(9):1302−11.

[158] Hofman P. PD-L1 immunohistochemistry for non-small cell lung carcinoma: which strategy should be adopted? Expert Rev Mol Diagn 2017;17(12):1097−108.

[159] Liu Y, et al. Heterogeneity of PD-L1 expression among the different histological components and metastatic lymph nodes in patients with resected lung adenosqua-mous carcinoma. Clin Lung Cancer 2018;19(4):e421−30.

[160] Ilié M, et al. Detection of PD-L1 in circulating tumor cells and white blood cells from patients with advanced non-small-cell lung cancer. Ann Oncol 2018;29(1):193−9.

[161] Guibert N, et al. PD-L1 expression in circulating tumor cells of advanced non-small cell lung cancer patients treated with nivolumab. Lung Cancer 2018;120:108−12.

[162] Nicolazzo C, et al. Monitoring PD-L1 positive circulating tumor cells in non-small cell lung cancer patients treated with the PD-1 inhibitor Nivolumab. Sci Rep 2016;6:31726.

[163] Satelli A, et al. Potential role of nuclear PD-L1 expression in cell-surface vimentin positive circulating tumor cells as a prognostic marker in cancer patients. Sci Rep 2016;6:28910.

Further reading

Demetri GD, Paz-Ares L, Farago AF, Liu SV, Chawla SP, Tosi D, et al. Efficacy and safety of entrectinib in patients with NTRK fusion-positive tumours: pooled analysis of STARTRK-2, STARTRK-1, and ALKA-372-001. Ann Oncol 2018;29:ix175. Available from: https://doi.org/10.1093/annonc/mdy483.003.

Marcus L, Lemery SJ, Keegan P, Pazdur R. FDA approval summary: pembrolizumab for the treatment of microsatellite instability-high solid tumors. Clin Cancer Res 2019;25 (13):3753. Available from: https://doi.org/10.1158/1078-0432.ccr-18-4070.

Yoshino T, Pentheroudakis G, Mishima S, Overman MJ, Yeh KH, Baba E, et al. JSCO—ESMO—ASCO—JSMO—TOS: international expert consensus recommendations for tumour-agnostic treatments in patients with solid tumours with microsatellite instabil-ity or NTRK fusions. Ann Oncol 2020. Available from: https://doi.org/10.1016/j.annonc.2020.03.299.

SUBCHAPTER 6.5

Circulating RNAs (miRNA, lncRNA, etc): from the enrichment to the characterization

D. Fanale[1], C. Brando[1], A. Fiorino[1], E. Pedone[1], A. Perez[1], M. Bono[1], N. Barraco[1], M. La Mantia[1], M. Del Re[2], A. Russo[1] and V. Bazan[3]

[1]Department of Surgical, Oncological, and Oral Sciences, University of Palermo, Palermo, Italy
[2]Unit of Clinical Pharmacology and Pharmacogenetics, Department of Clinical and Experimental Medicine, University of Pisa, Pisa, Italy
[3]Department of Biomedicine, Neuroscience and Advanced Diagnostics (Bi.N.D.), University of Palermo, Palermo, Italy

Learning objectives

By the end of the chapter, the reader will
*   have gained an overview of noncoding RNAs as potential biomarkers;
*   have learned the differences between various types of noncoding RNAs;
*   have discovered the role of noncoding RNAs in cancer and carcinogenesis.

6.5.1 Introduction

Large-scale genome sequencing has indicated that only 2% of the genome encodes for proteins, while more than 90% of the total genome is actively transcribed, but not uncodifying, and is known as noncoding RNA (ncRNA) [164−166].

These findings have changed the traditional view of RNA as an intermediary between DNA and proteins and demonstrate that most of the genome, long considered "evolutionary junk," encodes functional RNA species that differ in length and function [167].

Palazzo et al. [168] estimated that 99% of the RNA contained in mammalian cells is composed of ncRNA [164,169,170].

Based on length, ncRNAs can be distinguished into small-ncRNAs, if they consist of 18 to 200 nucleotides, and long-ncRNAs, if they consist of more than 200 nucleotides [170,171].

Based on function, however, we distinguish two different groups. The first one,07 housekeeping ncRNAs, includes ribosomal RNAs (rRNAs), transfer RNAs (t.,v./RNAs), small nucleolar RNAs (snoRNAs), and small nuclear RNAs (snRNA). The second one is regulatory ncRNA, key regulatory RNA molecules, that include microRNAs (miRNAs),

Table 6.5.1 Classification of ncRNAs.

Housekeeping ncRNAs	Regulatory ncRNAs	
	Small ncRNAs	Long ncRNA
Transfer RNAs (tRNAs)	Small interfering RNAs (siRNAs)	Long ncRNAs (lncRNAs)
Ribosomal RNAs (rRNAs)	PIWI-interacting RNAs (piRNAs)	Circular RNAs (circRNAs)
Small nucleolar RNAs (snoRNAs) Small nuclear RNAs (snRNAs)	microRNAs (miRNAs)	

PIWI-interacting RNAs (piRNAs), long noncoding RNAs (lncRNAs), and circular RNAs [170,172] (Table 6.5.1).

These circulating RNAs recently emerged in liquid biopsy as potential new biomarkers in several diseases and particularly in cancer, maybe secreted and persist in the biofluids in stable forms, as "free," associated with exosomes or packaged in extracellular vesicles [171,173]. They are minimally invasive, participate in numerous biological processes, and are aberrantly expressed under abnormal or pathological conditions [170].

Noninvasive biomarkers in the liquid biopsy are therefore a better option because they are easier to sample, less painful, and cheaper. In addition, they can be used in early diagnosis, can better characterize the tumor and its heterogeneity, and, through repeated sampling, allow better and more dynamic monitoring of disease progression and modulation of treatment [174] (Fig. 6.5.1).

6.5.2 Housekeeping RNAs

The class of housekeeping RNAs consists of a wide range of RNA species, many of which have been recently discovered and studied, and which perform various functions in almost all physiological and pathological processes [167]. These ncRNAs are usually small (50−500 nucleotides), are abundantly and ubiquitously expressed in all cell types, and perform a variety of cellular functions [172].

snRNAs are molecules consisting of 150 nucleotides located in the nucleus of the cell. Because of their location, they are involved in mRNA processing, spliceosome assembly, and translation.

Alterations in the expression of snRNAs might be involved in oncogenic processes. Several cell-free snRNAs have been identified in liquid

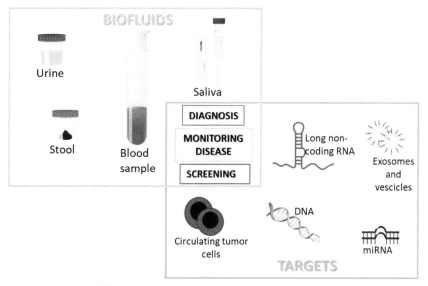

Figure 6.5.1 Role of liquid biopsy in the management of cancer using body fluids.

biopsy as cancer biomarkers, and the best-known family is indicated with the letter "U." In particular, snRNA U2 is increased in the blood of ovarian cancer patients and is associated with response to chemotherapy [175].

Another class of ncRNAs is snoRNAs, molecules of 60−300 nucleotides in length located in the nucleolus and involved in posttranscriptional rRNA editing and gene silencing [169]. snoRNAs can be divided into three classes based on sequence/structure/function: box C/D snoRNAs, box H/ACA snoRNAs, and scaRNAs, Cajal body-specific small RNAs primarily involved in posttranscriptional modification of small RNAs [169].

snoRNAs recognize and bind complementary sequences on target rRNAs and signal to proteins the exact base to be modified. snoRNAs and proteins form small nucleolar ribonucleoprotein complexes [169]. The role of snoRNAs in rRNA biogenesis has been well documented but, in the last few years, some studies highlighted other potential roles of this sncRNA in cellular regulation as well as other functions in cancer development and progression [169].

Several snoRNAs, detected by liquid biopsy in lung cancer, have been reported to be involved in the tumorigenesis and metastasis processes. Six snoRNAs are upregulated in the plasma of patients with nonsmall cell lung cancer (NSCLC) [176−180]. In another study, Mannor et al. [181] detected 22 snoRNAs deregulated in tumor-initiating cells of NSCLC, particularly with the expression of two snoRNAs (snoRA3 and snoRA42) being inversely associated with survival of NSCLC patients.

Approximately 4%—10% of all cellular RNAs are tRNAs, small noncoding ribonucleic acids folded into a trefoil structure of about 70—90 nucleotides involved in the transfer of amino acids to the site of protein synthesis. tRNAs originate from a precursor tRNA, which is an initial transcription product of RNA polymerase III (Pol III), and their synthesis is regulated by several oncogenes and tumor suppressor genes [182]. Ras and c-myc promote Pol III RNA transcription, while Rb and p53 inhibit Pol III RNA transcription, causing severe dysregulation of tRNA levels in a wide range of cancers [182,183].

The tRNA transcripts undergo enzymatic splicing, leading to the formation of new sncRNA species, such as tRNA-derived small RNAs (tsRNAs). Based on the length and cleavage site of the tRNA, tsRNAs can be divided into two broad categories: tRNA-derived stress-induced RNAs (tiRNAs) and tRNA-derived fragments (tRFs) [182].

tiRNAs are molecules approximately 31—40 nucleotides in length and are produced by cleavage of mature tRNAs at the central anticodon site. They are also called half tRNAs and are generated primarily by exposure to stress, such as oxidative stress, hypoxia, and viral infection [164].

tRFs, on the other hand, are 14—30 nucleotide length ncRNA molecules that originate from mature tRNAs or primary tRNA transcripts, having a 5′ phosphate and a 3′ hydroxyl group. Based on this, tRFs are classified as tRF-5 and tRF-3 if they originate from cleavage of the 5′ and 3′ ends of mature tRNAs, respectively, and tRF-1 if it originates from cleavage of the 3′ end of primary tRNA transcripts [169,184].

tRFs have been shown to contribute to carcinogenesis and are involved in cancer cell proliferation, metastasis, progression, and survival, representing potential diagnostic and prognostic biomarkers in different cancers, including gastric, liver, breast, and prostate cancer [184,185].

6.5.3 Regulatory ncRNAs

Regulatory ncRNAs are key molecules in the regulation of gene expression at the epigenetic, transcriptional, and posttranscriptional levels. They are a very large group of RNAs that include sncRNAs and lncRNAs [172].

The largest known class of ncRNAs are piRNAs, small RNA molecules with a length of about 26—31 nucleotides implicated in various physiological and pathological processes at the transcriptional or posttranscriptional level, including in transposon silencing, epigenetic regulation, and germline development [186,187]. After transcription, primary piRNA transcripts (prepiRNAs) are processed into mature piRNAs through two mechanisms: either through the primary synthesis mechanism, in which transcripts are generated from clusters of piRNA genes that bind to the PIWI protein, form the piRNA + PIWI complex, migrate into the nucleus and block transcription of the target gene, or through a mechanism, called "ping-pong," in which piRNAs, by binding to AGO3 or AUB proteins, cut an RNA sequence that will serve as a substrate for a new functional piRNA molecule [188].

piRNAs have been found in isolated exosomes but also freely circulating in serum, plasma, and saliva, resulting in promising noninvasive biomarkers with high sensitivity and specificity [189].

PIWI proteins and piRNAs are expressed in various human cancers, resulting in deregulated expression in different tumor tissues. Abundant dysregulated piRNAs play key roles in cancer cell proliferation, apoptosis, and metastasis, are strongly correlated with cancer cell malignant phenotype and clinical stage and may become potential prognostic and diagnostic biomarkers during cancer development. Deregulated expression of piRNAs has been reported in several human cancers, including gastric, bladder, breast, colorectal, and lung cancer [186,187,190−193]. Furthermore, the inhibition of piRNA expression can inhibit cancer development [194].

A new class of investigated RNAs is represented by lncRNAs, a heterogeneous group of transcripts that modulate many cellular functions [195] and act as key regulators (oncogenes or oncosopressors) in several malignancy processes [196]. With a total length of >200 nucleotides, they are localized in the nucleus, regulate the target gene expression, mainly through cis-regulation or trans-regulation, and modulate the mRNA processing, protein biogenesis, chromatin remodeling, carcinogenesis, and metastasis [197,198].

Of the more than 60,000 identified lncRNAs, only a few have been attributed a well-defined and clear function, and the different types of lncRNAs have been divided, based on their relative positions to protein-coding genes, into sense lncRNA, antisense lncRNA, bidirectional lncRNA, intron lncRNA, intergenic lncRNA, and enhancer lncRNA [197,199−201]. Of these, most are located in the nucleus, others play crucial roles in the cytoplasm and can be transferred to neighboring cells through exosomes, resulting from tissue/cell-specific [197].

Several tumors showed aberrant expression of lncRNAs, which are considered important regulators of cancer progression, since they are involved in cell proliferation, migration, invasion, epithelial-mesenchymal transition, apoptosis, and also in anticancer drug resistance mechanisms [197,202]. LncRNAs are implicated in some oncogenic pathways, such as those driven by p53, NF-κB, PI3K/AKT, Notch, Wnt/βcatenin, and HIF1α, suggesting that some lncRNAs are potential biomarkers for the diagnosis and prognosis of malignant tumors [203].

The last class of RNA discovered as potential biomarkers in liquid biopsy is represented by circular RNAs (circRNAs), socalled because of their circular shape in which the 3′ and 5′ ends are covalently linked [204].

This circular shape makes them biologically stable and resistant to RNases, thus resulting in useful biomarkers for the diagnosis and treatment of diseases [205].

Biologically, circRNAs regulate gene expression by acting as sponges and competitively binding miRNAs [173]. They can function as miRNA reservoirs, aid in miRNA transport, or suppress the binding of miRNAs to target genes [206]. In addition, circRNAs can also bind to RNA-binding proteins (RBPs), acting as protein sponges [207].

Although the interaction between circRNAs, miRNAs, and RBPs is complex and not fully understood, circRNAs have been studied in a wide range of diseases, including several types of cancers [208].

Most studies on the cancer-associated circRNAs focused on tissue biopsies and cell lines, but some works evaluated circRNAs only in body fluids [209−212], first detecting dysregulated circRNAs in tumor tissues, and then comparing their expression levels also in body fluids [213−218].

Although circRNAs are very stable and are detected in various body fluids, to date, their role as diagnostic biomarkers in liquid biopsy is mainly assessed on blood-derived samples. Only a few studies evaluated circRNAs in other body fluids, such as saliva and urine [219].

miRNAs are the best characterized and the most studied class of circulating RNAs [220]. miRNAs are small molecules (19−22 nucleotides) of noncoding RNA that regulate gene expression at the posttranscriptional level through binding to complementary nucleotides in the 3′ untranslated region (UTR) of messenger RNA (mRNA) target [221].

miRNAs can negatively regulate the gene expression posttranslationally by recognizing and binding to sequences within the 3′ UTR of target mRNAs, causing direct mRNA degradation or translational silencing [221,222]. A single miRNA can target multiple mRNAs and one mRNA can also be targeted by multiple miRNAs. However, some miRNAs can also positively regulate gene expression by promoting translation [164].

As regulators of gene expression, miRNAs play an important role in the initiation and development of many cancers, acting as oncogenes, called "Oncomirs," or as oncosuppressors, involved in cell proliferation, apoptosis, metastasis, and drug resistance [223−226]. Their critical roles in cancer find evident in a complex network of interactions underlying the relationship between miRNA regulation and the cancer hallmarks. As a consequence, the aberrant expression of miRNAs can affect a multitude of transcripts and different cancer-related signaling pathways [221,227−230].

The expression and function of normal miRNAs can be deregulated in cancer. Studies of miRNAs from cancer patients showed broad stability in various tissues and body fluids. MiRNAs, due to their small size and inability to be degraded by RNase, are generally stable in blood and other body fluids and are secreted by malignant cells into exosomes or stabilized and packaged with other proteins, including argonaute-2, high-density lipoprotein and other RNA-binding proteins [231]. Furthermore, the synergy of two or more miRNAs demonstrated great clinical value, supporting the development of combinatorial miRNA signatures (multiplex miRNA panels) as clinical tools.

However, the measurement of circ-miRNAs has poor consistency and reproducibility, due in part to low RNA concentration, short miRNA sequence length, high sequence similarity among miRNA family members, and sequence length variations. In addition, miRNAs are not disease- or organ-specific, and this makes them poorly adaptable as clinically relevant cancer diagnostic biomarkers [232].

Several other factors can affect miRNA measurement results. These include technical and nontechnical factors such as sample type, storage conditions, RNA isolation method, the physical state of the donor, and measurement platform [171].

As regards the miRNA measurement techniques, it is still quite difficult to accurately measure miRNAs, especially circulating miRNAs, although several platforms exist, including quantitative polymerase chain reaction (qPCR), microarrays, and next-generation sequencing (NGS) [171,233].

The qPCR platform, which relies on reverse transcription of miRNAs into complementary DNA, followed by polymerase chain reaction, is used to measure specific miRNAs and to validate miRNA profiling results obtained from platforms such as microarrays and NGS. It has high sensitivity, and ease of use but low throughput [171,234,235].

Microarray platforms are a commonly used hybridization-based method for miRNA profiling from cells and tissues. This is an inexpensive and high-throughput method but it may not be suitable for the analysis of circulating miRNAs due to the low concentration of RNAs in body fluids [171,236].

Finally, the NGS method is an emerging technology that has recently become the method of choice for circulating RNA analysis, allowing the identification of novel miRNAs and discrimination of isomiR. However, NGS-based methods require support from bioinformaticians for data analysis, sequencing library construction processes, and sequence biases arising from the sequencing library construction process [171].

To date, the search for informative diagnostic and prognostic markers based on noninvasive or minimally invasive techniques is at the forefront of many disease-oriented studies. Although the factors and mechanisms which may influence the spectrum of circulating RNA are not fully understood, several pieces of evidence suggest that circulating RNA, particularly miRNAs, may play an important role in disease diagnosis [171].

6.5.4 Key points

- only 2% of the mammalian genome encodes mRNAs, while the vast majority is transcribed, largely as long and short nonprotein-coding RNAs (ncRNAs);
- ncRNAs are a heterogeneous group of molecules, different in function and length, that represent more than 90% of the genome;
- circulating RNAs have recently emerged in liquid biopsy as potential new biomarkers in several diseases;
- ncRNAs are divided based on length (small and long noncoding RNAs) and function (housekeeping ncRNAs and regulatory ncRNAs);
- ncRNAs play different functions in physiological and pathological processes;
- RNA molecules, circulating freely in body fluids or packaged in microvesicles, have all the characteristics of ideal biomarkers (high stability under storage and handling conditions, possibility to be sampled multiple times for monitoring);
- Among all ncRNAs, miRNAs have been the most studied for their involvement in oncogenesis and their potential use as diagnostic and prognostic markers, but also as therapeutic means;
- The development of new sequencing technologies, such as next-generation sequencing, has made it possible to identify, quantify, and study different types of RNAs.

6.5.5 Expert opinion

- In the era of precision medicine, the search for non-invasive biomarkers is constantly increasing.

  Although several studies have shown that specific altered miRNA profiles in different types of cancers detected in blood and other body fluids can be used as candidate biomarkers, the application of circulating miRNAs for diagnosis, prognosis, and followup is still challenging.

  The limitations are still many. Indeed, the markers that have been identified have poor specificity and diagnostic reproducibility, and miRNA detection methods and data processing need better standardization of analytical methods before miRNAs are used for diagnostic purposes, both in the preprocessing and normalization procedures.

  Thus, there is still a need to minimize experimental or technical variations and ensure that results are reliable.

- Despite, the demonstrated stability of lncRNAs in serum it remains uncertain their binding and which signature is linked to precise clinical settings of neoplastic diseases. Studies about lncRNAs variations show that many serum RNAs have precise biological changes, demonstrating that they are not likely to be products of random degradation.

  Investigating what specific scars might characterize tumors could make the use of lncRNAs as biomarkers.

Acknowledgments

M. La Mantia, C. Brando, A. Fiorino, E. Pedone, and V. Gristina contributed to the current work under the Doctoral Program in Experimental Oncology and Surgery, University of Palermo.

References

[164] Chen H, Xu Z, Liu D. Small non-coding RNA and colorectal cancer. J Cell Mol Med 2019;23(5):3050−7.
[165] Fanale D, Barraco N, Listì A, Bazan V, Russo A. Non-coding RNAs functioning in colorectal cancer stem cells. Adv Exp Med Biol 2016;937:93−108.
[166] Russo A, Incorvaia L, Del Re M, Malapelle U, Capoluongo E, Gristina V, et al. The molecular profiling of solid tumors by liquid biopsy: a position paper of the AIOM-SIAPEC-IAP-SIBioC-SIC-SIF Italian Scientific Societies. ESMO Open 2021;6(3):100164.
[167] Taft RJ, Pang KC, Mercer TR, Dinger M, Mattick JS. Non-coding RNAs: regulators of disease. J Pathol 2010;220(2):126−39.
[168] Palazzo AF, Lee ES. Non-coding RNA: what is functional and what is junk? Front Genet 2015;6:2.

[169] Romano G, Veneziano D, Acunzo M, Croce CM. Small non-coding RNA and cancer. Carcinogenesis. 2017;38(5):485−91.

[170] Umu SU, Langseth H, Bucher-Johannessen C, Fromm B, Keller A, Meese E, et al. A comprehensive profile of circulating RNAs in human serum. RNA Biol 2018;15 (2):242−50.

[171] Lee I, Baxter D, Lee MY, Scherler K, Wang K. The importance of standardization on analyzing circulating RNA. Mol Diagn Ther 2017;21(3):259−68.

[172] Zhang P, Wu W, Chen Q, Chen M. Non-coding RNAs and their integrated networks. J Integr Bioinform 2019;16(3).

[173] Fanale D, Taverna S, Russo A, Bazan V. Circular RNA in exosomes. Adv Exp Med Biol 2018;1087:109−17.

[174] Pardini B, Sabo AA, Birolo G, Calin GA. Noncoding RNAs in extracellular fluids as cancer biomarkers: the new frontier of liquid biopsies. Cancers (Basel) 2019;11(8).

[175] Kuhlmann JD, Baraniskin A, Hahn SA, Mosel F, Bredemeier M, Wimberger P, et al. Circulating U2 small nuclear RNA fragments as a novel diagnostic tool for patients with epithelial ovarian cancer. Clin Chem 2014;60(1):206−13.

[176] Liao J, Yu L, Mei Y, Guarnera M, Shen J, Li R, et al. Small nucleolar RNA signatures as biomarkers for non-small-cell lung cancer. Mol Cancer 2010;9:198.

[177] Passiglia F, Galvano A, Castiglia M, Incorvaia L, Calò V, Listì A, et al. Monitoring blood biomarkers to predict nivolumab effectiveness in NSCLC patients. Ther Adv Med Oncol 2019;11 1758835919839928.

[178] Bian B, Fanale D, Dusetti N, Roque J, Pastor S, Chretien AS, et al. Prognostic significance of circulating PD-1, PD-L1, pan-BTN3As, BTN3A1 and BTLA in patients with pancreatic adenocarcinoma. Oncoimmunology 2019;8(4):e1561120.

[179] Incorvaia L, Fanale D, Badalamenti G, Barraco N, Bono M, Corsini LR, et al. Programmed death ligand 1 (PD-L1) as a predictive biomarker for pembrolizumab therapy in patients with advanced non-small-cell lung cancer (NSCLC). Adv Ther 2019;36(10):2600−17.

[180] Passiglia F, Rizzo S, Rolfo C, Galvano A, Bronte E, Incorvaia L, et al. Metastatic site location influences the diagnostic accuracy of ctDNA EGFR- mutation testing in NSCLC patients: a pooled analysis. Curr Cancer Drug Targets 2018;18(7):697−705.

[181] Mannoor K, Shen J, Liao J, Liu Z, Jiang F. Small nucleolar RNA signatures of lung tumor-initiating cells. Mol Cancer 2014;13:104.

[182] Huang SQ, Sun B, Xiong ZP, Shu Y, Zhou HH, Zhang W, et al. The dysregulation of tRNAs and tRNA derivatives in cancer. J Exp Clin Cancer Res 2018;37(1):101.

[183] Santos M, Fidalgo A, Varanda AS, Oliveira C, Santos MAS. tRNA deregulation and its consequences in cancer. Trends Mol Med 2019;25(10):853−65.

[184] Yu X, Xie Y, Zhang S, Song X, Xiao B, Yan Z. tRNA-derived fragments: mechanisms underlying their regulation of gene expression and potential applications as therapeutic targets in cancers and virus infections. Theranostics 2021;11(1):461−9.

[185] Fanale D, Incorvaia L, Badalamenti G, De Luca I, Algeri L, Bonasera A, et al. Prognostic role of plasma PD-1, PD-L1, pan-BTN3As and BTN3A1 in patients affected by metastatic gastrointestinal stromal tumors: can immune checkpoints act as a sentinel for short-term survival? Cancers (Basel) 2021;13(9).

[186] Liu Y, Dou M, Song X, Dong Y, Liu S, Liu H, et al. The emerging role of the piRNA/piwi complex in cancer. Mol Cancer 2019;18(1):123.

[187] Chalbatani GM, Dana H, Memari F, Gharagozlou E, Ashjaei S, Kheirandish P, et al. Biological function and molecular mechanism of piRNA in cancer. Pract Lab Med 2019;13:e00113.

[188] Assumpção CB, Calcagno DQ, Araújo TM, Santos SE, Santos Â, Riggins GJ, et al. The role of piRNA and its potential clinical implications in cancer. Epigenomics. 2015;7(6):975−84.

[189] Ferrero G, Cordero F, Tarallo S, Arigoni M, Riccardo F, Gallo G, et al. Small non-coding RNA profiling in human biofluids and surrogate tissues from healthy individuals: description of the diverse and most represented species. Oncotarget 2018;9(3):3097−111.

[190] Russo A, Incorvaia L, Malapelle U, Del Re M, Capoluongo E, Vincenzi B, et al. The tumor-agnostic treatment for patients with solid tumors: a position paper on behalf of the AIOM- SIAPEC/IAP-SIBioC-SIF Italian Scientific Societies. Crit Rev Oncol Hematol 2021;165:103436.

[191] Leto G, Incorvaia L, Flandina C, Ancona C, Fulfaro F, Crescimanno M, et al. Clinical impact of cystatin C/Cathepsin L and follistatin/activin a systems in breast cancer progression: a preliminary report. Cancer Invest 2016;34(9):415−23.

[192] Incorvaia L, Badalamenti G, Rinaldi G, Iovanna JL, Olive D, Swayden M, et al. Can the plasma PD-1 levels predict the presence and efficiency of tumor-infiltrating lymphocytes in patients with metastatic melanoma? Ther Adv Med Oncol 2019;11 1758835919848872.

[193] Galvano A, Taverna S, Badalamenti G, Incorvaia L, Castiglia M, Barraco N, et al. Detection of RAS mutations in circulating tumor DNA: a new weapon in an old war against colorectal cancer. A systematic review of literature and *meta*-analysis. Ther Adv Med Oncol 2019;11 1758835919874653.

[194] Cheng J, Guo JM, Xiao BX, Miao Y, Jiang Z, Zhou H, et al. piRNA, the new non-coding RNA, is aberrantly expressed in human cancer cells. Clin Chim Acta 2011;412(17−18):1621−5.

[195] Ghafouri-Fard S, Taheri M. Long non-coding RNA signature in gastric cancer. Exp Mol Pathol 2020;113:104365.

[196] Qiao X, Liu J, Zhu L, Song R, Zhong M, Guo Y. Long noncoding RNA CEBPA-DT promotes cisplatin chemo-resistance through CEBPA/BCL2 mediated apoptosis in oral squamous cellular cancer. Int J Med Sci 2021;18(16):3728−37.

[197] Chi Y, Wang D, Wang J, Yu W, Yang J. Long non-coding RNA in the pathogenesis of cancers. Cells 2019;8(9).

[198] Sanchez Calle A, Kawamura Y, Yamamoto Y, Takeshita F, Ochiya T. Emerging roles of long non-coding RNA in cancer. Cancer Sci 2018;109(7):2093−100.

[199] Wang KC, Chang HY. Molecular mechanisms of long noncoding RNAs. Mol Cell 2011;43(6):904−14.

[200] Badalamenti G, Barraco N, Incorvaia L, Galvano A, Fanale D, Cabibi D, et al. Are Long noncoding RNAs New potential biomarkers in gastrointestinal stromal tumors (GISTs)? the role of H19 and MALAT1. J Oncol 2019;2019:5458717.

[201] Incorvaia L, Badalamenti G, Rini G, Arcara C, Fricano S, Sferrazza C, et al. MMP-2, MMP-9 and activin A blood levels in patients with breast cancer or prostate cancer metastatic to the bone. Anticancer Res 2007;27(3B):1519−25.

[202] Fanale D, Castiglia M, Bazan V, Russo A. Involvement of non-coding RNAs in chemo- and radioresistance of colorectal cancer. Adv Exp Med Biol 2016;937:207−28.

[203] Peng WX, Koirala P, Mo YY. LncRNA-mediated regulation of cell signaling in cancer. Oncogene. 2017;36(41):5661−7.

[204] Ebbesen KK, Kjems J, Hansen TB. Circular RNAs: identification, biogenesis and function. Biochim Biophys Acta 2016;1859(1):163−8.

[205] Huang A, Zheng H, Wu Z, Chen M, Huang Y. Circular RNA-protein interactions: functions, mechanisms, and identification. Theranostics. 2020;10(8):3503−17.

[206] Ragan C, Goodall GJ, Shirokikh NE, Preiss T. Insights into the biogenesis and potential functions of exonic circular RNA. Sci Rep 2019;9(1):2048.

[207] Schneider T, Hung LH, Schreiner S, Starke S, Eckhof H, Rossbach O, et al. CircRNA-protein complexes: IMP3 protein component defines subfamily of circRNPs. Sci Rep 2016;6:31313.

[208] Wang Y, Liu J, Ma J, Sun T, Zhou Q, Wang W, et al. Exosomal circRNAs: bio-genesis, effect and application in human diseases. Mol Cancer 2019;18(1):116.
[209] Fan CM, Wang JP, Tang YY, Zhao J, He SY, Xiong F, et al. circMAN1A2 could serve as a novel serum biomarker for malignant tumors. Cancer Sci 2019;110(7):2180−8.
[210] Tang W, Fu K, Sun H, Rong D, Wang H, Cao H. CircRNA microarray profiling identifies a novel circulating biomarker for detection of gastric cancer. Mol Cancer 2018;17(1):137.
[211] Yin WB, Yan MG, Fang X, Guo JJ, Xiong W, Zhang RP. Circulating circular RNA hsa_circ_0001785 acts as a diagnostic biomarker for breast cancer detection. Clin Chim Acta 2018;487:363−8.
[212] Xu H, Gong Z, Shen Y, Fang Y, Zhong S. Circular RNA expression in extracellular vesicles isolated from serum of patients with endometrial cancer. Epigenomics. 2018;10(2):187−97.
[213] Shuai M, Hong J, Huang D, Zhang X, Tian Y. Upregulation of circRNA_0000285 serves as a prognostic biomarker for nasopharyngeal carcinoma and is involved in radiosensitivity. Oncol Lett 2018;16(5):6495−501.
[214] Zhang X, Zhou H, Jing W, Luo P, Qiu S, Liu X, et al. The circular RNA hsa_-circ_0001445 regulates the proliferation and migration of hepatocellular carcinoma and may serve as a diagnostic biomarker. Dis Markers 2018;2018:3073467.
[215] Li Z, Zhou Y, Yang G, He S, Qiu X, Zhang L, et al. Using circular RNA SMARCA5 as a potential novel biomarker for hepatocellular carcinoma. Clin Chim Acta 2019;492:37−44.
[216] Hang D, Zhou J, Qin N, Zhou W, Ma H, Jin G, et al. A novel plasma circular RNA circFARSA is a potential biomarker for non-small cell lung cancer. Cancer Med 2018;7(6):2783−91.
[217] Liu XX, Yang YE, Liu X, Zhang MY, Li R, Yin YH, et al. A two-circular RNA signature as a noninvasive diagnostic biomarker for lung adenocarcinoma. J Transl Med 2019;17(1):50.
[218] Tumminello FM, Badalamenti G, Incorvaia L, Fulfaro F, D'Amico C, Leto G. Serum interleukin-6 in patients with metastatic bone disease: correlation with cystatin C. Med Oncol 2009;26(1):10−15.
[219] Chen X, Chen RX, Wei WS, Li YH, Feng ZH, Tan L, et al. PRMT5 circular RNA promotes metastasis of urothelial carcinoma of the bladder through sponging MIR-30c to induce epithelial-mesenchymal transition. Clin Cancer Res 2018;24(24):6319−30.
[220] Cabibi D, Caruso S, Bazan V, Castiglia M, Bronte G, Ingrao S, et al. Analysis of tissue and circulating microRNA expression during metaplastic transformation of the esophagus. Oncotarget. 2016;7(30):47821−30.
[221] Incorvaia L, Fanale D, Badalamenti G, Brando C, Bono M, De Luca I, et al. A "Lymphocyte MicroRNA Signature" as predictive biomarker of immunotherapy response and plasma PD-1/PD-L1 expression levels in patients with metastatic renal cell carcinoma: pointing towards epigenetic reprogramming. Cancers (Basel) 2020;12(11).
[222] Taverna S, Cammarata G, Colomba P, Sciarrino S, Zizzo C, Francofonte D, et al. Pompe disease: pathogenesis, molecular genetics and diagnosis. Aging (Albany NY) 2020;12(15):15856−74.
[223] Caruso S, Bazan V, Rolfo C, Insalaco L, Fanale D, Bronte G, et al. MicroRNAs in colorectal cancer stem cells: new regulators of cancer stemness? Oncogenesis. 2012;1:e32.
[224] Amodeo V, Bazan V, Fanale D, Insalaco L, Caruso S, Cicero G, et al. Effects of anti-miR-182 on TSP-1 expression in human colon cancer cells: there is a sense in antisense? Expert Opin Ther Targets 2013;17(11):1249−61.
[225] Rolfo C, Fanale D, Hong DS, Tsimberidou AM, Piha-Paul SA, Pauwels P, et al. Impact of microRNAs in resistance to chemotherapy and novel targeted agents in non-small cell lung cancer. Curr Pharm Biotechnol 2014;15(5):475−85.

[226] Fanale D, Amodeo V, Bazan V, Insalaco L, Incorvaia L, Barraco N, et al. Can the microRNA expression profile help to identify novel targets for zoledronic acid in breast cancer? Oncotarget 2016;7(20):29321—32.

[227] Bronte F, Bronte G, Fanale D, Caruso S, Bronte E, Bavetta MG, et al. HepatomiRNoma: the proposal of a new network of targets for diagnosis, prognosis and therapy in hepatocellular carcinoma. Crit Rev Oncol Hematol 2016;97:312—21.

[228] Rizzo S, Cangemi A, Galvano A, Fanale D, Buscemi S, Ciaccio M, et al. Analysis of miRNA expression profile induced by short term starvation in breast cancer cells treated with doxorubicin. Oncotarget. 2017;8(42):71924—32.

[229] Leto G, Incorvaia L, Badalamenti G, Tumminello FM, Gebbia N, Flandina C, et al. Activin A circulating levels in patients with bone metastasis from breast or prostate cancer. Clin Exp Metastasis 2006;23(2):117—22.

[230] Tumminello FM, Badalamenti G, Fulfaro F, Incorvaia L, Crescimanno M, Flandina C, et al. Serum follistatin in patients with prostate cancer metastatic to the bone. Clin Exp Metastasis 2010;27(8):549—55.

[231] Corsini LR, Bronte G, Terrasi M, Amodeo V, Fanale D, Fiorentino E, et al. The role of microRNAs in cancer: diagnostic and prognostic biomarkers and targets of therapies. Expert Opin Ther Targets 2012;16(Suppl 2):S103—9.

[232] Shigeyasu K, Toden S, Zumwalt TJ, Okugawa Y, Goel A. Emerging role of MicroRNAs as liquid biopsy biomarkers in gastrointestinal cancers. Clin Cancer Res 2017;23(10):2391—9.

[233] Incorvaia L, Fanale D, Bono M, et al. BRCA1/2 pathogenic variants in triple-negative vs luminal-like breast cancers: genotype-phenotype correlation in a cohort of 531 patients. Ther Adv Med Oncol 2020;12. Available from: https://doi.org/10.1177/1758835920975326. 1758835920975326. Published 2020 Dec 16.

[234] Pisapia P, Pepe F, Gristina V, et al. A narrative review on the implementation of liquid biopsy as a diagnostic tool in thoracic tumors during the COVID-19 pandemic. Mediastinum 2021;5:27. Available from: https://doi.org/10.21037/med-21-9. Published 2021 Sep 25.

[235] Incorvaia L, Russo A, Cinieri S. The molecular tumor board: a tool for the governance of precision oncology in the real world. Tumori 2021. Available from: https://doi.org/10.1177/03008916211062266. Epub ahead of print. PMID: 34918610.

[236] Russo A, Incorvaia L, Capoluongo E, Tagliaferri P, Galvano A, Del Re M, et al. The challenge of the molecular tumor board empowerment in clinical oncology practice: a position paper on behalf of the AIOM-SIAPEC/IAP-SIBioC-SIC-SIF-SIGU-SIRM Italian Scientific Societies. Crit Rev Oncol Hematol 2022;169:103567. Available from: https://doi.org/10.1016/j.critrevonc.2021.103567. Epub 2021 Dec 8. PMID: 34896250.

Further reading

Anfossi S, Babayan A, Pantel K, et al. Clinical utility of circulating non-coding RNAs—an update. Nat Rev Clin Oncol 2018;15:541—63. Available from: https://doi.org/10.1038/s41571-018-0035-x.

Li J, Guo S, Sun Z, Fu Y. Noncoding RNAs in drug resistance of gastrointestinal stromal tumor. Front Cell Dev Biol 2022;10:808591. Available from: https://doi.org/10.3389/fcell.2022.808591. Available from: 35174150. PMCID: PMC8841737.

Palazzo AF, Lee ES. Non-coding RNA: what is functional and what is junk? Front Genet. 2015;6:2. Available from: https://doi.org/10.3389/fgene.2015.00002. Published 2015 Jan 26.

Pardini B, Sabo AA, Birolo G, et al. Noncoding RNAs in extracellular fluids as cancer biomarkers: the new frontier of liquid biopsies. Cancers 2019;11:1170. Available from: https://doi.org/10.3390/cancers11081170.

Preethi KA, Selvakumar SC, Ross K, et al. Liquid biopsy: exosomal microRNAs as novel diagnostic and prognostic biomarkers in cancer. Mol Cancer. 2022;21(1):54. Available from: https://doi.org/10.1186/s12943-022-01525-9. Available from: 35172817. PMCID: PMC8848669.

Umu SU, Langseth H, Bucher-Johannessen C, et al. A comprehensive profile of circulating RNAs in human serum RNA Biol. 2018;15(2):242—50. Available from: https://doi.org/10.1080/15476286.2017.1403003. Epub 2017 Dec 8. Available from: 29219730. PMCID: PMC5798962.

SUBCHAPTER 6.6

Cell-free/circulating tumor DNA profiling: from next-generation sequencing-based to digital polymerase chain reaction-based methods

A. Perez[1,*], C. Brando[1,*], M. La Mantia[1,*], V. Gristina[1], A. Galvano[1], L. Incorvaia[1],

G. Badalamenti[1], A. Giordano[2], E. Capoluongo[3,4], A. Russo[1], U. Malapelle[5,†] and V. Bazan[6,†]

[1]Department of Surgical, Oncological, and Oral Sciences, University of Palermo, Palermo, Italy
[2]Sbarro Institute for Cancer Research and Molecular Medicine, and Center for Biotechnology, College of Science and Technology, Temple University, Philadelphia, PA, United States
[3]Department of Molecular Medicine and Medical Biotechnology, University of Naples Federico II, Naples, Italy
[4]CEINGE, Advances Biotecnologies, Naples, Italy
[5]Department of Public Health, University of Naples Federico II, Naples, Italy
[6]Department of Biomedicine, Neuroscience and Advanced Diagnostics (Bi.N.D.), University of Palermo, Palermo, Italy

Learning objectives

By the end of the chapter, the reader will:

- Have learned the basic concepts of the main molecular biology techniques;
- Have learned the most feasible next-generation sequencing (NGS) and digital PCR (dPCR) platforms for cell-free/circulating tumor DNA genomic profiling;
- Be able to distinguish between targeted and untargeted methods;
- Be confident with the most appropriate approach to use.

* Perez A, Brando C, La Mantia M. should be considered equally cofirst authors.
† Bazan V and Malapelle U. should be considered equally colast authors.

6.6.1 Introduction

In recent years, the recognition in clinical practice of the liquid biopsy-based diagnostic utility has completely revolutionized the standard old vision, primarily relying on tissue-based approaches (Fig. 6.6.1). Among circulating nucleic acids, cell-free DNA (cfDNA) represents one of the most investigated newborn biomarkers for its highly recognized role in highlighting molecular tumor dynamics for both prognostic and predictive purposes. In particular, cfDNA molecules originate from several cell types through a wide broad of physiological and pathological mechanisms, from the hematopoietic system to a multitude of programmed biological events such as apoptosis, senescence, ferroptosis, NETosis, phagocytosis, and necrosis, along with secretion and several other processes still under investigation [237,238]. The circulating tumor DNA (ctDNA), the tumor-derived fraction of its parental cfDNA, has acquired a growing interest in the last decade for its enormous potential as a poorly invasive tumor biomarker in several oncological settings. Nowadays, the biggest efforts in the use of ctDNA in oncology have been primarily concentrated on the detection of peculiar tumor-linked mutations in body fluids of cancer patients [239]. The detection of specific predictive and/or prognostic genetic variants on already known genes has the potential to be used in many oncological settings as early cancer detection, verifying the origin of

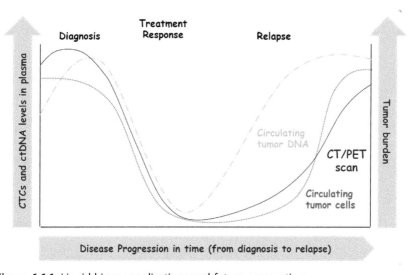

Figure 6.6.1 Liquid biopsy applications and future perspectives.

the tumor, tracking and predicting molecular response to treatment, assessing secondary resistance mechanisms, or to detect minimal residual disease (MRD) [240−243]. Moreover, several published studies have deeply highlighted a high concordance between circulating DNA and its tissue counterpart, while analyzing single-nucleotide genetic alterations in KRAS, NRAS, PIK3CA, BRAF, and EGFR hotspot genes in colorectal [244−248], lung [238,249−252], and breast cancer patients [245,253−256]. This assessment is strongly correlated, limited, and consequently dependent, on the wide variable amount of ctDNA released into the bloodstream from every tumor, as well as the high variability among patients with the same cancer type. Therefore, the choice of the proper body fluid strongly depends on the success of the liquid biopsy-based diagnostic approach. Indeed, to overcome this limit, nonblood sources have become of great interest, for their higher sensitivity if compared with peripheral blood, as the use of cerebrospinal fluid for adult and pediatric primary brain tumors, or saliva, sputum, or pleural effusions for aerodigestive tract tumors [255,257,258]. In recent years, given the outstanding potential of liquid biopsy in daily clinical practice, an important advancement has been carried out for both the detection and analysis of ctDNA. In particular, its evaluation is particularly challenging due to a plethora of ctDNA intrinsic characteristics. Indeed, ctDNA, if compared to its tissutal counterpart, is highly shredded and its total amount can be often extremely low with respect to background nucleic acids as the fraction of tumoral DNA content could be <0.01% with respect to total physiologically released DNA [259−261]. Moreover, the lack of standardization methods currently limits the feasibility of ctDNA approaches in which the single-molecule sensitivity needs to be accompanied by an accurate specificity [262]. The technologies for ctDNA profiling can be mainly classified into targeted and untargeted approaches for their abilities to target already known recurrent mutations or discover still unknown genetic variants. Indeed, targeted approaches are characterized by a high sensitivity and specificity, a more cost-effective rate, and, therefore, can be very useful in specific disease settings as the evaluation of minimal residual disease, the detection of disease recurrence or to track resistance molecular mechanisms [263]. Contrariwise, untargeted approaches even if less sensitive with respect to the targeted counterpart, are useful for highlighting new genetic and genomic alterations such as copy number variations (CNV). This chapter will focus on the targeted and untargeted technologies suitable and currently in use for ctDNA profiling.

6.6.2 Targeted next-generation sequencing methods

Nowadays, the use of targeted NGS-based approaches allows a massive and parallel analysis of thousands of short DNA regions, thus covering whole or entire coding regions of several genes at the same time. One of the first steps in library preparation is through different approaches: direct amplification (amplicon-based methods) or hybridization (hybrid capture-based methods) of the regions of interest. The first amplicon-based method is strictly dependent and limited by the DNA fragments' length and often requires, for enrichment, several amplification reactions to cover more regions of a gene and consequently a variable amount of DNA. Contrariwise, hybridization capture methods rely on small RNA probes complementary to targeted regions allowing the detection of both single nucleotide variants and large rearrangements using a lower DNA amount [264]. One of the main issues while working with ultra-deep sequencing methods and with very low sensitivities, is the possibility, at any step during the NGS, of introducing several polymerase errors, losing thus reliability to confidently call rare variants, and consequently discriminating true from false called variants. Indeed, a casual error rate between 0.1 and 1.5%, has been estimated depending on the specific NGS platform considered [265,266]. Currently, the introduction and use of molecular tags (so-called barcodes), before any amplification steps, enabled the direct count and discrimination of original DNA molecules from PCR duplicates allowing thus a more accurate mutation screening and profile with increased sensitivity [267−269]. The use of DNA barcoding is common but at the same time strongly differentiates and drives the progress of NGS methodologies. Tagged-amplicon deep sequencing (Tam-Seq) is an amplicon-based method combining a library preparation using barcoded primers with statistically-based analysis algorithms. In particular, after the first amplification step of the target regions, sequencing adaptors and sample barcodes are end-attached to the newborn amplified sequences. This approach is suitable for the detection and quantification of tumor hotspot variants as along with whole coding regions of specific genes of interest. Interestingly, Tam-Seq has been recently enhanced (so-called enhanced Tam-Seq, eTam-Seq) with strategies allowing not only the amplification of highly fragmented DNA, such as ctDNA, but more reliability to confidently call rare SNV, indels, and also CNVs, reducing thus unbiased error rates. The study conducted by Gale et al., starting from an optimal amount of DNA, demonstrated the ability of eTam-Seq, with high performance, to detect low frequency circulating somatic alterations with 94% mutations ranging between 0.25%–0.33% allele fraction, down to 0.02%, with a per-base specificity of 99.9997% for EGFR mutations in cfDNA [270,271]. To further improve NGS sensitivity, Safe-Sequencing System (Safe-SeqS) has been introduced. Its enhancement consists of the assignment of a unique identifier (UID) to each template molecule before the amplification steps to amplify each uniquely tagged template to set up homogeneous UID sequence families. Consequently, are considered mutant only those UID families fragments harboring the same mutation in more than 95% of them. Therefore, this method, by lowering amplification and sequencing bias, can detect genetic variants, in cfDNA fragments, at a very low allelic fraction in many clinical settings, from real-time MRD monitoring to molecular profiling in advanced stages [272,273]. Furthermore, a recent study conducted by Tie et al., in three independent cohorts of nonmetastatic CRCs demonstrated a median variant allele frequency of 0.046% (interquartile range, 0.010%–0.191%) by applying Safe-SeqS technology for the molecular profiling of cfDNA [274]. An improvement of Safe-SeqS technology is Duplex Sequencing in which a double-stranded barcoded adapter is used and attached to double-stranded DNA. Consequently, after sequencing, are considered mutated at high confidence only those fragments not only containing the duplex adaptors but in which there is a consensus between both strands. As deeply demonstrated by multiple studies on plasma cfDNA this enhanced method, if used with the hybrid capture-based approach as targeting enrichment, allowed to reach a 0.1% allelic fraction detection along with high sensitivity and specificity, for diagnosis and monitoring of stage IV colorectal cancer [275,276]. Furthermore, to further facilitate the discernment of true and false positive calls, Targeted Error Correction sequencing (TEC-Seq) has been evaluated. Interestingly, this approach combines different barcoding strategies. Indeed, in addition to "exogenous DNA barcodes," before any amplification, TEC-Seq uses specific genome positions of paired-end sequenced fragments, as "endogenous barcodes." This strategy allows following every single DNA fragment through the sequencing process [277,278]. The application of this method to several solid tumors highlighted strong reliability in detecting genetic variants at an allelic frequency ranging between 0.1%–0.2% with a 97.4% sensitivity performance, and, surprisingly, without identification of false positives. Single primer extension is an amplicon-based approach specifically introduced by QIAGEN for its own-designed targeted panels using a unique gene-specific primer for amplification of each target region and adapters containing 12 base pair sequence barcodes (Unique Molecular Index). The incorporation of a unique sequence on each DNA fragment allows true variant alleles, present in the original sample, to be distinguished from possible mistakes introduced during NGS steps, from library preparation to target enrichment and sequencing. Therefore, since each nucleic acid is uniquely tagged with a specific molecular barcode, duplicate reads, as well as PCR errors, can be filtered out by bioinformatics software with a high level of accuracy reducing thus the rate of false-positive variant calls. This method had been deeply applied at diagnosis for the detection of actionable genetic alteration but has been also improved for the detection of CNVs at high sensitivity [267,279,280]. Cancer Personalized Profiling by Deep Sequencing (CAPP-Seq) is a hybridization-based NGS method specifically designed for the evaluation of ctDNA at an extremely low concentration. This method relies on biotinylated DNA oligonucleotides, so-called "selectors," complementary to specific genomic regions. Hybridization is then followed by deep sequencing allowing the detection of single nucleotide variants, rearrangements, and CNVs [281,282]. Interestingly, CAPP-Seq allowed the detection of genetic variants at an allelic frequency of 0.02% in ctDNA of NSCLC patients with a 96% specificity. CAPP-Seq has been then improved (so-called iDES-enhanced CAPP-Seq) through the combination of the CAPP-Seq method, duplex barcoding sequencing technology, and a computational tool iDES (Integrated Digital Error Suppression), specifically designed to correct sequencing or PCR system errors. In fact, iDES-enhanced CAPP-Seq demonstrated a high sensibility and specificity ($>99.99\%$) as a detection method for EGFR mutation at an allele fraction lower than 0.004% in NSCLC plasma samples by using a DNA input of only 32 ng. Furthermore, as assumed by Newman et al., this combinatorial strategy could reach a theoretical detection of very low rate mutant allele frequency down to 0.00025% [283].

6.6.3 Untargeted next-generation sequencing methods

Untargeted approaches can be mainly classified into whole exome and whole genome sequencing (WES, WGS). They are characterized by less sensitivity, compared to targeted methods, in the detection of genetic alterations on cfDNA [284,285]. Their sensitivity is estimated to be 5%−10% when compared to a sensitivity of 0.1% of the targeted sequencing approaches, making the detection of rare events particularly difficult, especially at early detection and/or MRD [284]. Moreover, the technologies to support whole exome/ genome sequencing are very expensive to be afforded by routine laboratories and often require the support of expert bioinformatics able to analyze and filter the enormous amount of data generated by a single analysis. Therefore, all these characteristics make this approach still difficult to apply to everyday clinical practice. Contrariwise, an untargeted approach can unravel the presence of new genetic and genomic alterations above all at diagnosis for initial molecular profiling and, considering the intratumoral molecular variability, to discover potential pharmaceutical targets and highlight drug resistance [286]. WES is a suitable approach to studying coding regions and splicing sites of genes containing most of the relevant genetic alteration disease-related along with promoters, UTRs, and noncoding DNA regions containing for example miRNA genetic information [287]. Moreover, if compared to the targeted counterpart is characterized by low coverage and sensitivity making particularly difficult the detection of less representative and unexpected variants. The feasibility of WES for mutational analysis of cfDNA has been primarily associated with advanced tumor stages and consequently a ctDNA allele fraction greater than 5%. The utility of using a whole-exome approach in clinical practice has been interestingly exploited by Murtaza et al., who performed a sequencing analysis on ctDNA extracted from serial plasma samples to highlight molecular evolution in several advanced cancer patients receiving systemic therapy [288]. The analysis performed on samples at high ctDNA percentage (5%−55%) identifies interesting genetic alterations responsible for treatment failure, demonstrating that cfDNA could be a feasible specimen to describe intratumoral heterogeneity. Contrariwise, WGS analysis on ctDNA could be helpful to identify copy number variation as well as methylation profiles, and fragmentation patterns. Several groups worldwide have exploited the use of a WGS approach to identify CNVs from plasma-derived ctDNA even if the high costs and timing. In particular, Heitzer and colleagues developed the Plasma-seq genome-wide sequencing approach using an Illumina MiSeq instrument detecting CNVs in plasma cfDNA, from patients with prostate cancer, at high specificity ($>80\%$) and ctDNA allele fraction greater than 10% [289]. Interestingly, PARE (personalized analysis of rearranged ends) and WGS analysis developed by Leary et al. have demonstrated suitability to detect genomic rearrangements even at ctDNA fraction as low as 0.001% with respect to total cfDNA amount [290]. Moreover, several studies in ctDNA of breast and colorectal cancer patients suggested the possibility to detect with a sensitivity higher than 90% and a specificity greater than 99% even a single copy of rearrangement without errors in calling false positives [291]. Recently, PARE analysis has been considered to predict 12 months postsurgery relapse in patients affected by gastric tumor through the detection of rearranged loci in cfDNA of gastric tumor patients [292].

6.6.4 Droplet digital polymerase chain reaction methods

Among PCR-based approaches, digital PCR, firstly coined by Vogelstein and Kinzler in the late 1990s, represents the third generation PCR, a robust upgrade of the highly widespread real-time PCR technique [293]. The study and introduction of this high-performing technique arise from the urgent need to detect genetic variants present at a very low allele fraction in a wide background of wild-type fragments. This approach allows obtaining qualitative and absolute quantitative information at the same time for the same sample. Indeed, the contrary of real-time PCR, absolute quantification of DNA and RNA can be achieved without the need for a standard curve because the target is ideally partitioned into thousand separated reactions, allowing thus the detection of small percentage variations and the quantification of rare variants as copies/μL. The most spread dPCR platforms use droplet-based technologies, in particular droplet digital PCR (ddPCR), and BEAMing (beads, emulsions, amplification, and magnetics). These approaches find in cancer research and diagnostics one of the main fields of application, in particular in the evaluation of circulating tumor DNA (ctDNA), also generally known as the term "liquid biopsy."

ddPCR is a breakthrough technology firstly developed to provide ultrasensitive and absolute quantification of targets present at a very low-abundance fraction such as somatic variants that cannot be detected with standard PCR-based methods [294]. The ddPCR is based on small water-in-oil microemulsion droplets in which target and background DNA are randomly distributed among thousand droplets. The template is subsequently amplified by end-point PCR in every single droplet that behaves as a separate reaction microenvironment. Each microreactor could ideally guest or not a copy of a mutated or wild-type DNA target that is detected by using fluorescent probes and scored as positive or negative based on any fluorescence emission. Therefore, the count of positive and negative droplets adjusted by Poisson statistical analysis can provide absolute quantitation of the target sequence. Therefore, given its ultrasensitive characteristics, ddPCR is particularly suitable in all those settings that require the detection of a small amount of nucleic acid input for a wide range of applications, from rare sequence detection to copy number variation and gene expression analysis of rare targets. Consequently, the ddPCR approach is characterized by a very deep sensitivity in detecting genetic alterations present at a very low fraction at a quite affordable cost for most laboratories but the main disadvantage remains the limit of screening only known and a limited number of variants in a single reaction [286,295,296]. As previously mentioned, another droplet-based platform used is BEAMing, based on beads, emulsion, amplification, and magnetics. BEAMing has been firstly described in 2003 by Vogelstein's team and then commercialized late in 2014 [293]. This technology uses small water-oil droplets, with a $3-10\,\mu m$ diameter, representing an independent PCR reaction environment. In particular, each emulsion droplet contains the reaction mixture needed for a PCR reaction along with primer-coated magnetic beads. Indeed, this technology uses primers that are covalently bound, via streptavidin-biotin, to the surface of magnetic beads. Following the PCR amplification steps, the droplets emulsion is then disrupted to release the magnetic beads containing or not the amplified target attached on their surface which in turn will be magnetically purified, and then fluorescent probes will be attached. A flow cytometer will then analyze the fluorescent signal deriving from probes binding, identifying wild-type or mutated DNA targets and also the mutant fraction with respect to total wild-type DNA [112]. This massively parallel PCR platform demonstrated high sensitivity (0.01%−0.001%) and specificity (100%), with a concordance of 90% between tissue and ctDNA, in several types of cancer [297−300]. Unfortunately, given its complicated workflow and the poorly affordable costs of the equipment, BEAMing use is still less spread in clinical practice [285,301].

6.6.5 Key points

1. The recognition in clinical practice of the liquid biopsy-based diagnostic utility has completely revolutionized the standard old vision, primarily relying on tissue-based approaches;
2. The technologies for ctDNA profiling can be mainly classified into targeted and untargeted approaches for their abilities to target already known recurrent mutations or discover still unknown genetic variants;
3. The use of NGS-based approaches allows a massive and parallel analysis of thousands of short DNA regions, thus covering whole or entire coding regions of several genes at the same time;

4. Amplicon-based methods or hybrid capture-based methods represent the most spread approaches used for NGS analysis;

5. The most spread dPCR platforms use droplet-based technologies, in particular droplet digital PCR (ddPCR), and BEAMing (beads, emulsions, amplification, and magnetics).

6.6.6 Expert opinion

- Given the crucial role of tumor molecular profiling in guiding personalized treatment planning, molecular biology techniques have become an integral and fundamental part of clinical oncology routine;
- Since the first biological sequencing methods, the most modern technological advancements for application in molecular oncology are Next-Generation Sequencing and ddPCR;
- Molecular biology approaches provide a feasible tool for clinical and translational research in oncology due to their high sensibility and specificity;
- Nowadays, oncology cannot be anymore considered a separate field; it is perfectly complemented by molecular biology.

Acknowledgments

M. La Mantia, C. Brando, and V. Gristina contributed to the current work under the Doctoral Program in Experimental Oncology and Surgery, University of Palermo.

References

[237] Keller L, Belloum Y, Wikman H, Pantel K. Clinical relevance of blood-based ctDNA analysis: mutation detection and beyond. Br J Cancer 2021;124(2):345−58.

[238] Russo A, Incorvaia L, Del Re M, Malapelle U, Capoluongo E, Gristina V, et al. The molecular profiling of solid tumors by liquid biopsy: a position paper of the AIOM-SIAPEC-IAP-SIBioC-SIC-SIF Italian Scientific Societies. ESMO open 2021;6(3):100164.

[239] Malapelle U, Pepe F, Pisapia P, Sgariglia R, Nacchio M, Barberis M, et al. TargetPlex FFPE-Direct DNA Library Preparation Kit for SiRe NGS panel: an international performance evaluation study. J Clin Pathol 2021.

[240] Wan JCM, Massie C, Garcia-Corbacho J, Mouliere F, Brenton JD, Caldas C, et al. Liquid biopsies come of age: towards implementation of circulating tumour DNA. Nat Rev Cancer 2017;17(4):223−38.

[241] Heitzer E, Haque IS, Roberts CES, Speicher MR. Current and future perspectives of liquid biopsies in genomics-driven oncology. Nat Rev Genet 2019.

[242] Passiglia F, Galvano A, Gristina V, Barraco N, Castiglia M, Perez A, et al. Is there any place for PD-1/CTLA-4 inhibitors combination in the first-line treatment of advanced NSCLC?-a trial-level *meta*-analysis in PD-L1 selected subgroups. Transl Lung Cancer Res 2021;10(7):3106−19.

[243] Gristina V, La Mantia M, Galvano A, Cutaia S, Barraco N, Castiglia M, et al. Non-small cell lung cancer harboring concurrent egfr genomic alterations: a systematic review and critical appraisal of the double dilemma. J Mol Pathol 2021;2 (2):173−96. Available from: https://www.mdpi.com/2673-5261/2/2/16.

[244] Kuo Y-B, Chen J-S, Fan C-W, Li Y-S, Chan E-C. Comparison of KRAS mutation analysis of primary tumors and matched circulating cell-free DNA in plasmas of patients with colorectal cancer. Clin Chim Acta 2014;433:284−9.

[245] Bettegowda C, Sausen M, Leary RJ, Kinde I, Wang Y, Agrawal N, et al. Detection of circulating tumor DNA in early- and late-stage human malignancies. Sci Transl Med 2014.

[246] Galvano A, Gristina V, Malapelle U, Pisapia P, Pepe F, Barraco N, et al. The prognostic impact of tumor mutational burden (TMB) in the first-line management of advanced non-oncogene addicted non-small-cell lung cancer (NSCLC): a systematic review and meta-analysis of randomized controlled trials. ESMO Open 2021;6(3). Available from: https://doi.org/10.1016/j.esmoop.2021.100124.

[247] Pisapia P, Pepe F, Gristina V, La Mantia M, Francomano V, Russo G, et al. A narrative review on the implementation of liquid biopsy as a diagnostic tool in thoracic tumors during the COVID-19 pandemic Mediastinum 2021;5. Mediastinum [Internet]. 2021. Available from: https://med.amegroups.com/article/view/6433.

[248] Galvano A, Taverna S, Badalamenti G, Incorvaia L, Castiglia M, Barraco N, et al. Detection of RAS mutations in circulating tumor DNA: a new weapon in an old war against colorectal cancer. A systematic review of literature and meta-analysis. Ther Adv Med Oncol 2019;11. 1758835919874653.

[249] Fernandez-Cuesta L, Perdomo S, Avogbe PH, Leblay N, Delhomme TM, Gaborieau V, et al. Identification of circulating tumor DNA for the early detection of small-cell lung cancer. EBioMedicine. 2016;10:117−23.

[250] Yuan H, Zhu Z-Z, Lu Y, Liu F, Zhang W, Huang G, et al. A modified extraction method of circulating free DNA for epidermal growth factor receptor mutation analysis. Yonsei Med J 2012;53(1):132−7.

[251] Gristina V, La Mantia M, Iacono F, Galvano A, Russo A, Bazan V. The emerging therapeutic landscape of ALK inhibitors in non-small cell lung cancer. Pharm (Basel) 2020;13(12).

[252] Beretta G, Capoluongo E, Danesi R, Del Re M, Fassan M, Giuffrè G, et al. Raccomandazioni 2020 per l'esecuzione di Test Molecolari su Biopsia Liquida in Oncologia, 2020.

[253] Oshiro C, Kagara N, Naoi Y, Shimoda M, Shimomura A, Maruyama N, et al. PIK3CA mutations in serum DNA are predictive of recurrence in primary breast cancer patients. Breast Cancer Res Treat 2015;150(2):299−307.

[254] Incorvaia L, Fanale D, Bono M, Calò V, Fiorino A, Brando C, et al. BRCA1/2 pathogenic variants in triple-negative vs luminal-like breast cancers: genotype-phenotype correlation in a cohort of 531 patients. Ther Adv Med Oncol 2020;12. 1758835920975326.

[255] Russo A, Incorvaia L, Malapelle U, Del Re M, Capoluongo E, Vincenzi B, et al. The tumor-agnostic treatment for patients with solid tumors: a position paper on behalf of the AIOM- SIAPEC/IAP-SIBioC-SIF Italian Scientific Societies. Crit Rev Oncol Hematol 2021;165:103436.

[256] Pisapia P, Pepe F, Baggi A, Barberis M, Galvano A, Gristina V, et al. Next generation diagnostic algorithm in non-small cell lung cancer predictive molecular pathology: the KWAY Italian multicenter cost evaluation study. Crit Rev Oncol Hematol 2021;169:103525.

[257] Boire A, Brandsma D, Brastianos PK, Le Rhun E, Ahluwalia M, Junck L, et al. Liquid biopsy in central nervous system metastases: a RANO review and proposals for clinical applications. Neuro Oncol 2019;21(5):571−84.

[258] Abbou SD, Shulman DS, DuBois SG, Crompton BD. Assessment of circulating tumor DNA in pediatric solid tumors: the promise of liquid biopsies. Pediatr Blood Cancer 2019;66(5):e27595.

[259] Cheng F, Su L, Qian C. Circulating tumor DNA: a promising biomarker in the liquid biopsy of cancer. Oncotarget 2016;7(30):48832−41.

[260] Badalamenti G, Barraco N, Incorvaia L, Galvano A, Fanale D, Cabibi D, et al. Are long noncoding RNAs new potential biomarkers in gastrointestinal stromal tumors (GISTs)? The role of H19 and MALAT1. J Oncol 2019;2019:5458717.

[261] Passiglia F, Rizzo S, Di Maio M, Galvano A, Badalamenti G, Listì A, et al. The diagnostic accuracy of circulating tumor DNA for the detection of EGFR-T790M mutation in NSCLC: a systematic review and *meta*-analysis. Sci Rep 2018;8(1):13379.

[262] Leto G, Incorvaia L, Flandina C, Ancona C, Fulfaro F, Crescimanno M, et al. Clinical impact of cystatin C/Cathepsin L and follistatin/activin a systems in breast cancer progression: a preliminary report. Cancer Invest 2016;34(9):415−23.

[263] Incorvaia L, Fanale D, Badalamenti G, Barraco N, Bono M, Corsini LR, et al. Programmed death ligand 1 (PD-L1) as a predictive biomarker for pembrolizumab therapy in patients with advanced non-small-cell lung cancer (NSCLC). Adv Ther 2019.

[264] Mamanova L, Coffey AJ, Scott CE, Kozarewa I, Turner EH, Kumar A, et al. Target-enrichment strategies for next-generation sequencing. Nat Methods 2010;7(2):111−18.

[265] Glenn TC. Field guide to next-generation DNA sequencers. Mol Ecol Resour 2011;11(5):759−69.

[266] Loman NJ, Misra RV, Dallman TJ, Constantinidou C, Gharbia SE, Wain J, et al. Performance comparison of benchtop high-throughput sequencing platforms. Nat Biotechnol 2012;30(5):434−9.

[267] Xu C, Nezami Ranjbar MR, Wu Z, DiCarlo J, Wang Y. Detecting very low allele fraction variants using targeted DNA sequencing and a novel molecular barcode-aware variant caller. BMC Genomics 2017;18(1):5.

[268] Galvano A, Castiglia M, Rizzo S, Silvestris N, Brunetti O, Vaccaro G, et al. Moving the target on the optimal adjuvant strategy for resected pancreatic cancers: a systematic review with *meta*-analysis. Cancers (Basel) 2020;12(3).

[269] Passiglia F, Bronte G, Bazan V, Natoli C, Rizzo S, Galvano A, et al. PD-L1 expression as predictive biomarker in patients with NSCLC: a pooled analysis. Oncotarget 2016;7(15):19738−47.

[270] Gale D, Lawson ARJ, Howarth K, Madi M, Durham B, Smalley S, et al. Development of a highly sensitive liquid biopsy platform to detect clinically-relevant cancer mutations at low allele fractions in cell-free DNA. PLoS One 2018;13(3):e0194630.

[271] Listì A, Barraco N, Bono M, Insalaco L, Castellana L, Cutaia S, et al. Immuno-targeted combinations in oncogene-addicted non-small cell lung cancer. Transl Cancer Res 2019;8(Suppl 1) Transl Cancer Res (Targeted ther non-small cell lung cancer a new era?) 2018; https://tcr.amegroups.com/article/view/24695.

[272] Kinde I, Wu J, Papadopoulos N, Kinzler KW, Vogelstein B. Detection and quantification of rare mutations with massively parallel sequencing. Proc Natl Acad Sci U S A 2011.

[273] Passiglia F, Galvano A, Rizzo S, Incorvaia L, Listì A, Bazan V, et al. Looking for the best immune-checkpoint inhibitor in pre-treated NSCLC patients: an indirect comparison between nivolumab, pembrolizumab and atezolizumab. Int J Cancer 2018.

[274] Tie J, Cohen JD, Lo SN, Wang Y, Li L, Christie M, et al. Prognostic significance of postsurgery circulating tumor DNA in nonmetastatic colorectal cancer: individual patient pooled analysis of three cohort studies. Int J cancer 2021;148(4):1014−26.

[275] Ren Y, Zhang Y, Wang D, Liu F, Fu Y, Xiang S, et al. SinoDuplex: an improved duplex sequencing approach to detect low-frequency variants in plasma cfDNA samples. Genomics Proteom Bioinforma 2020;18(1):81−90.

[276] Mallampati S, Zalles S, Duose DY, Hu PC, Medeiros LJ, Wistuba II, et al. Development and application of duplex sequencing strategy for cell-free DNA-based longitudinal monitoring of stage IV colorectal cancer. J Mol Diagn 2019;21(6):994—1009.

[277] Phallen J, Sausen M, Adleff V, Leal A, Hruban C, White J, et al. Direct detection of early-stage cancers using circulating tumor DNA. Sci Transl Med 2017;9(403).

[278] Massihnia D, Galvano A, Fanale D, Perez A, Castiglia M, Incorvaia L, et al. Triple negative breast cancer: shedding light onto the role of pi3k/akt/mtor pathway. Oncotarget 2016;7(37):60712—22.

[279] Peng Q, Xu C, Kim D, Lewis M, DiCarlo J, Wang Y. Targeted single primer enrichment sequencing with single end duplex-UMI. Sci Rep 2019;9(1):4810.

[280] Viailly P-J, Sater V, Viennot M, Bohers E, Vergne N, Berard C, et al. Improving high-resolution copy number variation analysis from next generation sequencing using unique molecular identifiers. BMC Bioinforma 2021;22(1):120.

[281] Newman AM, Bratman SV, To J, Wynne JF, Eclov NCW, Modlin LA, et al. An ultrasensitive method for quantitating circulating tumor DNA with broad patient coverage. Nat Med 2014;20(5):548—54.

[282] Bratman SV, Newman AM, Alizadeh AA, Diehn M. Potential clinical utility of ultrasensitive circulating tumor DNA detection with CAPP-Seq. Vol. 15. Expert Rev Mol diagnostics 2015;715—19.

[283] Newman AM, Lovejoy AF, Klass DM, Kurtz DM, Chabon JJ, Scherer F, et al. Integrated digital error suppression for improved detection of circulating tumor DNA. Nat Biotechnol 2016;34(5):547—55.

[284] Corcoran RB, Chabner BA. Application of cell-free DNA analysis to cancer treatment. N Engl J Med 2018.

[285] Russo A, Incorvaia L, Capoluongo E, Tagliaferri P, Galvano A, Del Re M, et al. The challenge of the molecular tumor board empowerment in clinical oncology practice: a position paper on behalf of the AIOM- SIAPEC/IAP-SIBioC-SIC-SIF-SIGU-SIRM Italian Scientific Societies. Crit Rev Oncol Hematol 2022;169:103567. Available from: https://doi.org/10.1016/j.critre-vonc.2021.103567. Available from: 34896250. Available from: https://www.sciencedirect.com/science/article/pii/S1040842821003541.

[286] Bos MK, Angus L, Nasserinejad K, Jager A, Jansen MPHM, Martens JWM, et al. Whole exome sequencing of cell-free DNA—a systematic review and Bayesian individual patient data meta-analysis. Cancer Treat Rev 2020;83:101951.

[287] Choi M, Scholl UI, Ji W, Liu T, Tikhonova IR, Zumbo P, et al. Genetic diagnosis by whole exome capture and massively parallel DNA sequencing. Proc Natl Acad Sci U S A 2009;106(45):19096—101.

[288] Murtaza M, Dawson S-J, Tsui DWY, Gale D, Forshew T, Piskorz AM, et al. Non-invasive analysis of acquired resistance to cancer therapy by sequencing of plasma DNA. Nature 2013;497(7447):108—12.

[289] Heitzer E, Ulz P, Belic J, Gutschi S, Quehenberger F, Fischereder K, et al. Tumor-associated copy number changes in the circulation of patients with prostate cancer identified through whole-genome sequencing. Genome Med 2013;5(4):30.

[290] Pareek CS, Smoczynski R, Tretyn A. Sequencing technologies and genome sequencing. J Appl Genet 2011;52(4):413—35.

[291] Leary RJ, Sausen M, Kinde I, Papadopoulos N, Carpten JD, Craig D, et al. Detection of chromosomal alterations in the circulation of cancer patients with whole-genome sequencing. Sci Transl Med 2012;4(162):162ra154.

[292] Kim Y-W, Kim Y-H, Song Y, Kim H-S, Sim HW, Poojan S, et al. Monitoring circulating tumor DNA by analyzing personalized cancer-specific rearrangements to detect recurrence in gastric cancer. Exp Mol Med 2019;51(8):1—10.

[293] Morley AA. Digital PCR: a brief history. Biomol Detect Quantif 2014;1(1):1−2.
[294] Kristensen LS, Hansen LL. PCR-based methods for detecting single-locus DNA methylation biomarkers in cancer diagnostics, prognostics, and response to treatment. Clin Chem 2009;55(8):1471−83.
[295] Gorgannezhad L, Umer M, Islam MN, Nguyen N-T, Shiddiky MJA. Circulating tumor DNA and liquid biopsy: opportunities, challenges, and recent advances in detection technologies. Lab Chip 2018;18(8):1174−96.
[296] Bohers E, Viailly P-J, Jardin F. cfDNA sequencing: technological approaches and bioinformatic issues. Pharmaceuticals (Basel) 2021;14(6).
[297] Diehl F, Schmidt K, Choti MA, Romans K, Goodman S, Li M, et al. Circulating mutant DNA to assess tumor dynamics. Nat Med 2008;14(9):985−90.
[298] Higgins MJ, Jelovac D, Barnathan E, Blair B, Slater S, Powers P, et al. Detection of tumor PIK3CA status in metastatic breast cancer using peripheral blood. Clin Cancer Res 2012;18(12):3462−9.
[299] Li M, Diehl F, Dressman D, Vogelstein B, Kinzler KW. BEAMing up for detection and quantification of rare sequence variants. Nat Methods 2006;3(2):95−7.
[300] Elazezy M, Joosse SA. Techniques of using circulating tumor DNA as a liquid biopsy component in cancer management. Comput Struct Biotechnol J 2018;16:370−8.
[301] Incorvaia L, Russo A, Cinieri S. The molecular tumor board: a tool for the governance of precision oncology in the real world. Tumori 2021. Available from: https://doi.org/10.1177/03008916211062266. Epub ahead of print. PMID: 34918610.

SUBCHAPTER 6.7

Standardization and quality assurance in liquid biopsy testing

A. Perez[1], M. Bono[1], N. Barraco[1], C. Brando[1], D. Cancelliere[1], A. Pivetti[1], A. Fiorino[1], E. Pedone[1], A. Giurintano[1], M. La Mantia[1], V. Gristina[1], A. Galvano[1], L. Incorvaia[1], G. Badalamenti[1], U. Malapelle[2,*], A. Russo[1,*] and V. Bazan[3,*]

[1]Department of Surgical, Oncological, and Oral Sciences, University of Palermo, Palermo, Italy
[2]Department of Public Health, University of Naples Federico II, Naples, Italy
[3]Department of Biomedicine, Neuroscience and Advanced Diagnostics (Bi.N.D.), University of Palermo, Palermo, Italy

Learning objectives

- To understand the high potential of liquid biopsy in better supporting clinical decisions;
- To understand the limits of cell-free DNA (cfDNA), CTCs, and Exosomes at a preanalytical level as a crucial step for liquid biopsy applications;
- To understand the current standardization initiatives and external quality assessment (EQAs) schemes to better improve liquid biopsy applications.

* Malapelle U, Russo A, Bazan V should be considered equally colast authors.

6.7.1 Introduction

The advent of precision medicine brings the promise that therapy can become personalized to each patient. Drug-targeted medicine depends heavily on advances in molecular biology and new genetic tools to accurately stratify and characterize patients. The promise of precision medicine becomes more achievable with the concept of liquid biopsies, an intelligent alternative to solid biopsies [302]. Indeed, liquid biopsy represents one of the most important steps forward through a completely personalized approach, especially in the oncology field. Indeed, in the last years, it is a more and more urgent need for a complete landscape of tumor heterogeneity to gain information about tumor molecular characteristics to better support clinical decisions [303–305]. The preanalytical phase is a crucial step of the analysis and plays an important role as the analytical one. Unfortunately, the lack of strong evidence of clinical utility and validity along with the lack of standardized methods, especially at a preclinical level, still limits the clinical success of such an approach [242,306–308]. This chapter will deeply discuss the limits linked to the preanalytical phase of the main components of liquid biopsy Cell-free DNA [(cfDNA), circulating tumor cells (CTCs), and exosomes] along with the most recent standardization initiatives aimed at improving the data and the reach of a worldwide consensus for diagnostic purposes in liquid biopsy (Fig. 6.7.1).

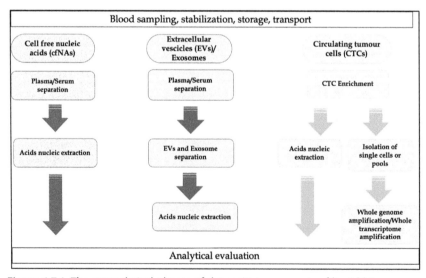

Figure 6.7.1 The preanalytical phases of the main components of liquid biopsy.

6.7.2 Cell-free DNA in liquid biopsy: preanalytical limitations

The human cells have always guarded our genetic and epigenetic information as well as cancer disease too. If we want to take a snapshot, it would be necessary to include the genome fragmented released in the circulatory system and different body fluids [309−311]. These cells are the sentinels of a possible deregulated genomic condition thanks to their fascinating state of stability. Indeed, the experts' opinions of the AIOM-SIAPEC, IAP, SIBIOC, and SIF Italian Scientific Societies members have defined these cells as ideal surrogate biomarkers of treatment response, easily discovered in cancer patients. In particular, the source of cell-free DNA (cfDNA) molecules and circulating tumor DNA (ctDNA), that is cfDNA shed from tumor cells through cell death mechanisms such as necrosis and apoptosis in biofluids, is represented by different sample types such as blood and its derivatives that include circulating tumor cells, circulating tumor RNA, microRNAs, platelets, extracellular vesicles, saliva, and urine, cerebrospinal fluid, or other fluids. The limitations to achieving information for diagnostic, prognostic, and predictive purposes in the preanalytical phase have been distinguished at biological and technological levels [312]. cfDNA molecules, fragments of the genome released into body fluids and circulation by human cells, are stable and maintain the distinctive genetic and epigenetic characteristics of the cells from which they originate. This feature makes cfDNA the ideal candidate for detecting and monitoring genomic alterations, particularly those present in solid tumors [313−315]. The characteristics of cfDNA depend on the type of mechanism by which it is released from the cell, several pathways appear to be involved but the main ones are apoptosis and necrosis. Apoptosis results in the presence of short cfDNA fragments (166−498 bp), whereas necrosis usually generates fragments larger than 10,000 bp [316,317]. cfDNA refers to the entire population of extracellular DNA, which includes both ctDNA, which usually represents the smallest fraction, and wild-type cfDNA, which is the most represented. The ratio of the percentage of the mutated allele to the wild type allele can be described as the mutant allele fraction. One of the most relevant features that significantly influences the quality, in terms of purity, integrity, and concentration, of cfDNA is the type of tube used for cfDNA biospecimen collection. EDTA tubes are the most widely used for cfDNA studies, both in diagnostics and research, as the DNase-inhibiting properties of EDTA are well known and certainly preferred over heparin tubes, which have an inhibitory effect on PCR. Due to the short half-life of cfDNA, estimated to be around 2.5 h [318,319], storage of blood at room temperature in EDTA tubes should not exceed 3 h, and plasma collection should be done as soon as possible after blood collection. However, it is not always possible to meet this time frame, especially when samples have to be transported from off-site sites to laboratories for analysis, it is not always possible to process the sample within 3 h. To meet these requirements, new blood collection tubes have been developed in recent years that stabilize samples for longer periods of time. These tubes can maintain the integrity of cfDNA samples because they contain a preservative agent that prevents lysis of peripheral blood cells, thus avoiding germline DNA contamination that results in tumor DNA dilution. To ensure clinically significant sensitivity, it is essential to qualify and quantify the isolated cfDNA [320−322]. Only then can the cfDNA yield be analyzed to identify biomarkers associated with targeted therapies. Regarding the technological matter, the frontiers to detect the common and uncommon pathogenetic variants, according to specific clinical contexts, are due to the same nature of the tumoral cells along with the body fluids, as little as 0.01% fraction of all circulating cells (minor allele frequency). High plasma input and methods with a low limit of detection (LoD) are necessary to sequence sufficient ctDNA concentrations to be able to detect low minor allele frequency. Moreover, the plethora of platforms used for molecular testing of cell-free nucleic acids (cfDNAs) is not validated. Despite that, real-time Polymerase chain reaction (Real-time PCR), Digital PCR (dPCR), Beads, Emulsion, Amplification, and Magnetics (BEAMing) followed by Next-Generation Sequencing (NGS) technologies, considering the low abundance and intermediate sensitivity of plasma ctDNA, share a high sensitivity. To date, Real-time PCR is the widespread system because it assures turnaround time and proper sensitivity at low costs. The presence of deregulated hotspot genes such as single-base mutations and small deletions can be detected by the Scorpion ARMS technology. This consists of a specific Scorpion primer-probe complex mechanism where a PCR primer covalently linked to a probe, which in turn interacts with a quencher, detects one or more mutated alleles of the target gene and an endogenous control gene. The quality and quantity of cfDNAs in the reaction are checked by the mixture of control oligonucleotides. A qualitative scale of the risk ratio between false positives (FP) and false negatives (FN) associated with real-time PCR is in equilibrium and it is possible to reach a LOD of 1%−5% [323−325]. Instead, the technological advancement of the classic PCR is the innovative approach of dPCR in which the cfDNA templates are split into thousands of single parallel PCR reactions. Indeed, the term "digital" has been used to indicate the exponential output of the PCR, historically linear, in dichotomic binary signal: presence (positive) or absence (negative) of the target sequence into bioreactions (droplets or chips) after amplification. The three types of dPCR are: droplet dPCR (ddPCR), BEAMing dPCR and solid digital PCR. In ddPCR and dsPCR systems, the cfNA templates are partitioned, according to Poisson distribution, into bioreactions represented by homogeneous droplets (~ 20.000) in an oil-water emulsion and wells (~ 12.000) on-chip (solid support), respectively. The second strategy attempts to overcome the performance limits of PCR amplification step in the emulsion. The fluorescence emission of small numbers of mutated in a background of wild-type alleles can provide qualitative and quantitative information, but the output must be interpreted through the expert eye of biologists/biotechnologists. This reproducible data performs better than Real-time PCR for detecting the know mutations, reaching a sensitivity limit of 0.1%-0.01%, but a qualitative scale of the risk ratio between FP and FN shows an imbalance toward FPs. The high-throughput NGS-based genotype methods are the last but not least approach. These fascinating technologies offer the opportunity to use different research kits to investigate in ctNAs multiple genes and multiple known and unknown alterations (single-nucleotide variant, in/dels, rearrangements) simultaneously. These sensitive NGS applications allowed a sensitivity of 0.1%-0.01% to be reached, but, as for dPCR system, a qualitative scale of the risk ratio between FP and FN shows an imbalance toward FPs [326,327].

6.7.3 The preanalytical phase of circulating tumor cells analysis

CTCs were described nearly a century ago and are defined as cells released into the bloodstream by the tumor tissue and are responsible for the development of metastases. However, the use of CTCs in clinical practice has been shown in several scenarios: screening, prognosis, and treatment response. However, clinical application is still limited. This is likely due to the lack of standardization methods and the presence of complications in the preanalytical phase. One of the most common difficulties related to the CTCs used is their fragility which leads to their degradation within hours [328,329]. In addition, in the early phase of the disease, the presence of CTCs can be rare and very low in frequency making isolation and enrichment difficult, two of the most common problems [330–332]. The first step in the preanalytical phase involves the collection of samples to perform CTCs analysis. Blood is most commonly collected in K3EDTA tubes, taking care to be quick in processing the samples to avoid cell lysis. However, one study compared K3EDTA tubes to tubes with preservatives such as CellSave (Veridex) and BCTs (Streck) before detection and counting of CTCs. In tubes with preservatives, CTCs were maintained for 4 days with higher recovery rates, in contrast for K3EDTA tubes the recovery rates were much lower [316,333,334]. Another group compared BCT tubes vs EDTA, citrate, and heparin tubes for the detection of CTCs. This article showed that BCT tubes performed higher in the detection of CTCs in patients with breast cancer [335,336]. Finally, Ilie M et al. in 2018 compared K3EDTA and BCT tubes in the enrichment of CTCs by size using the ISET system (Rarecells Diagnostics, Paris, France) in NSCLC patients. The researchers demonstrated that the CTCs count was stable, intact, and morphologically correct in BCT tubes at 24 and 48 h after blood sampling, compared with K3EDTA tubes [322]. The second critical step in the preanalytical phase is the isolation and enrichment of CTCs. To date, several techniques exploit the physical or biological properties of CTCs. Isolation of CTCs through physical properties is done by evaluating characteristics such as cell size, density, or deformability. In addition, an approach based on biological properties is exploitable. These properties are focused on cell surface markers, through positive selection with markers expressed on tumor cells, or negative selection with markers expressed on leukocytes [337–339]. Nowadays, enrichment with positive selection is the most widely used approach, while the only FDA-approved system is the CTCs counting by CellSearch (Menarini Silicon Biosystem, Bologna, Italy) in breast, colon, and prostate cancers [340–342]. Currently, research on CTCs focuses on the single-cell approach and molecular/genomic description of CTCs, which can provide a large contribution to detecting genomic mutations or subclonal variations specific to CTCs [343]. However, the single-cell approach is a laborious and difficult process and often requires a preenrichment step followed by single-cell sorting. However, this technique is not yet routine in clinical practice, and universal standardization would be needed to use the single-cell method in the analysis of CTCs to answer the various interrogations on tumor heterogeneity and evolution [314,344].

6.7.4 The preanalytical phase of exosomes analysis

A promising new target of LB is exosomes, extracellular endosomal vesicles of variable size (typically 30 to 100 nm), widely distributed through body fluids and easily accessible by blood, saliva, breast milk, or urine. These vesicles are bounded by a phospholipid bilayer and internally enclose various biomolecules including proteins, lipids, glycans, metabolites, and nucleic acids (DNA, mRNA, miRNA, lncRNA) [324,345,346]. In recent years, exosomes have emerged as an area of intense interest because they have been recognized as a powerful means of intercellular communication due to their ability to transfer biomolecules that affect various functions of target cells [347]. Most healthy cells produce exosomes under homeostatic conditions, but it has been shown that in tumor cells the release is significantly increased and that tumor-derived exosomes show a specific burden reflecting tumor type. Once these are taken up by target cells, they can influence their phenotype [348]. Because exosomes can be isolated from all body fluids, they are ideal candidates for noninvasive liquid biopsy analysis not only in the diagnosis of various diseases, but also in other biomedical fields, including drug delivery, cell-free vaccine development, and regenerative medicine. Nevertheless, studies of exosomes derived from body fluids represent a challenge, because isolated exosomes are derived from many different cell types and contain body fluid components, which must be removed during exosome isolation. Recent technological advances in exosome isolation and their analysis have not led to a universally accepted recommendation, and to date, the International Society for Extracellular Vesicles guidelines are still vague [349].

In recent decades, although there is no standardized method of exosome isolation, numerous techniques have been established [350]. Exosome characterization and isolation are technically challenging: preanalytical variables (sample collection, storage, transport, and exosome isolation) can affect exosome size, morphology, and yield, and postanalytical variables (data acquisition and analysis) vary widely [350].

6.7.4.1 Sample collection, storage, and processing

Pre-analytical steps represent an important source of variability in exosome analysis, already starting with the collection of fluid containing extracellular vesicles (EVs), which is a delicate step. The use of plasma is generally recommended, although serum has appropriate uses. Collected blood should be handled gently and processed quickly at room temperature. The use of a large needle is recommended for venipuncture to minimize shear forces and avoid endothelial stress and activation. Once the blood has been collected, the tubes should be placed vertically and may be inverted for proper mixing with the anticoagulant. Ca2 + chelation-based anticoagulants [sodium citrate or ethylenediaminetetraacetic acid (EDTA)] are commonly the ones used. Citrate tubes are generally preferred over EDTA, which can often affect EV measurements. The time between blood collection and centrifugation also has an impact on downstream analysis. EV may be stable for the first 30 min after extraction, but a long incubation before processing results in increased EV levels.

6.7.4.2 Exosomes isolation

Exosomes are heterogeneous nanometer-sized vesicles that differ in size, content, function, and origin, which makes isolation difficult [351]. So, the method used for isolation from body fluids is critical for the characterization of their cargo, and to date, most current isolation technologies are unable to completely separate exosomes from lipoproteins with similar biophysical characteristics causing poor exosomal purity compromising their quantification and functional [311]. To date, six classes of exosome separation strategies have been reported, including ultracentrifugation, ultrafiltration, size-exclusion chromatography, polymer precipitation, immunoaffinity-based approaches, and microfluidic techniques, and each of which has different advantages and disadvantages (Table 6.7.1).

6.7.4.2.1 Ultracentrifugation-based isolation techniques

Ultracentrifugation (UC) is considered the gold standard and is currently the most widely used isolation technique for exosome extraction and isolation. UC mainly collects components based on the differences in size and density of each component in the original solution [350]. Centrifugation time, centrifugal force, rotor type, and parameters all influence the yield and purity of target exosomes. There are two types of UC: analytical and preparative UC. Analytical UC is used to study the physicochemical properties of particulate materials and the molecular interactions of polymeric materials. In contrast, preparative UC is used to fractionate biological components and it is divided into two groups: differential UC and density gradient UC.

Differential UC also referred to as simple UC or pelletizing method, is the most commonly reported strategy for exosome separation (Fig. 6.7.2A). Isolation of exosomes by differential UC is quite simple and usually consists

Table 6.7.1 Comparison of current exosome isolation techniques.

Isolation methods Technique	Principle	Advantages	Disadvantages	References
Differential ultracentrifugation	Separation by density, size, and shape of the different components of a fluidic sample through cycles of increasing centrifugal force for a defined time.	• Simple procedure; • Low technical expertise; • Accessibility over time; • No or minimal sample pretreatment.	• Expensive appropriate instrumentation; • Highly time-consuming; • Lack of automation; • Contamination with protein aggregates, viruses, and other microvesicles; • Low purity; • Structural damage due to high speed; • Not suitable for small volumes.	PMID: 33061359 PMID: 32206116 PMID: 30604646
Density-gradient centrifugation	After centrifugation in a dense medium (sucrose, iodixanol) the molecules sediment in the position of the medium with a similar density	• High exosome purity; • Minimal contamination.	• Expensive appropriate instrumentation; • Highly time-consuming; • Lack of automation; • Highly qualified technicians; • Structural damage due to high speed; • Not suitable for small volumes.	PMID: 27009329 PMID: 18675270 PMID: 29662902
Ultrafiltration	Isolation is performed through the use of membrane filters with defined molecular weight or size exclusion limits	• Fast procedure; • Low cost; • High enrichment efficiency.	• Possible low efficiency due to membrane blockage and entrapment.	PMID: 30148165 PMID: 25834266

(Continued)

Table 6.7.1 (Continued)

Isolation methods Technique	Principle	Advantages	Disadvantages	References
Size-exclusion chromatography (SEC)	A porous stationary phase is used to elute particles based on their size	• Simple procedure; • Applicable for various types of fluids; • No or minimal sample pretreatment; • Saves time and labor; • Does not affect the structure and integrity of vesicles; • No or minimal sample loss; • High output.	• An exosome enrichment method is required.	PMID: 27640641 PMID: 29316942 PMID: 30987213
Polymer precipitation	Precipitation is induced by the use of a polymer (PEG) that alters the solubility of exosomes	• Simple procedure; • High output; • No specialized equipment is required; • Short analysis time; • Applicable to both small and large volumes.	• Possibility of coprecipitation of protein aggregates, other extracellular vesicles, and polymeric contaminants, which affect the purity of the preparation; • Precipitation matrices could affect the activities and biological characteristics of the exosome.	PMID: 27068479 PMID: 27009329 PMID: 30987213
Immunoaffinity-based approaches	Based on the specific binding of markers in the exosome to immobilized antibodies	• Simple and fast procedure; • High exosome purity; • Isolation of specific exosome subtypes; • No volume limits.	• High cost; • Possible damage to exosome structure; • Antibody specificity and quality; • Reduce exosome output.	PMID: 32899828 PMID: 28255367
Microfluidic-based exosome isolation	Through microfluidic devices, exosomes are isolated based on several principles including immunoaffinity, size, and density	• Efficient and fast processing; • High level of purity; • Easy automation and integration; • High portability.	• Moderate to low sampling capacity.	PMID: 28832692 PMID: 31666080

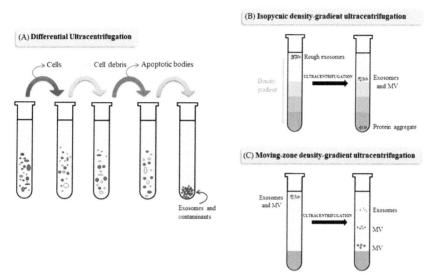

Figure 6.7.2 *Schematic representation of exosome isolation based on ultracentrifugation.* (A) Differential ultracentrifugation provides several cycles of centrifugation. After each cycle, pellets, which include cells, cellular debris, and apoptotic bodies, are removed, and the supernatant is collected for further centrifugation. After the last centrifugation, the supernatant is removed and pellets containing exosomes and contaminants (including protein aggregates, viruses, and microvesicles) are collected. (B) Density gradient isopycnic ultracentrifugation is performed by the addition of a biocompatible medium (iodoxinol, sucrose) with gradually decreasing densities from bottom to top into a tube. Next, the sample is placed at the top of the medium and a centrifuge cycle is performed. In this way, the different extracellular components reach a static position, called isopycnic, in the layer with the same density. However, this method does not separate substances, such as microvesicles, which have a similar density to exosomes. (C) Moving-zone gradient ultracentrifugation presents a top layer with a lower density than all solutes in the sample and a bottom layer with high density. After centrifugation, the solutes in the sample will be separated not only by density but also by mass/size.

of a series of centrifugation cycles of different centrifugal forces and duration to isolate exosomes based on their density and size differences from other components in a sample. For UC, the centrifugal force used typically ranges from ~100,000 to 120,000 × g. UC is primarily divided into two steps: first, a series of continuous centrifugations at low to medium speeds to remove dead cells, cellular debris, and large extracellular vesicles, followed by multiple cycles of centrifugation with higher strength and speeds to remove impurities and contaminants such as apoptotic bodies and protein aggregates for purified exosome isolation [352]. Centrifugation time, centrifugal force, rotor type, and parameters all affect the yield and purity of target exosomes. The technique is easy to apply and does not require technical expertise, but the resulting samples have low purity since several components (including exosomes, microvesicles, and protein aggregates) can precipitate at the bottom of the tube, compromising downstream analysis. Based on these observations and to improve these separation techniques, density gradient centrifugation, which separates particles by density, was developed.

In *density gradient UC*, separation of exosomes is achieved based on their size, mass, and density.

This technique uses a specific centrifuge tube that contains a preconstructed density gradient gradually decreasing from the bottom to the top, generally consisting of sucrose and iodixanol [352]. The sample is added to the top of the density gradient medium and subjected to prolonged centrifugation for a period of time. By applying a centrifugal force, extracellular components (apoptotic bodies and protein aggregates) including exosomes in the sample move across the density gradient toward the bottom. Each component then reaches a position in the layer of its density.

Through this method, the protein aggregates concentrate at the bottom of the centrifuge tube while the exosomes remain in the medium layer and can thus be easily recovered [350]. Although density gradient UC is limited by the narrow loading zone, it has an advantage in terms of higher exosome purity [352]. There are two types of density gradient UC, namely isopycnic UC and moving zone UC (Fig. 6.7.2B−C). In isopycnic UC, the separation of exosomes from other solutes depends solely on their difference in density. During centrifugation, the exosomes sediment along with the density gradient medium until they reach and maintain their position in the layer of the same density called the isopycnic position [350]. However, because isopycnic gradient UC depends on the difference in density between different solutes, it is not possible to separate substances that have similar buoyancy densities to exosomes, such as microvesicles. In contrast, in moving-zone UC, exosomes are separated based on their size and mass rather than density. Because the density of solutes, including exosomes, is greater than the density of the gradient medium, they will settle to the bottom of the tube. This technique allows the separation of extracellular vesicles with similar densities but different sizes.

6.7.4.2.2 Ultrafiltration and size exclusion chromatography
In these two techniques, exosome isolation is based on size differences between exosomes and other components in a biological sample. Specifically,

the separation principle of size exclusion chromatography uses an early stationary phase that separates macromolecules and particulates based on their size. Smaller molecules enter the pores while larger molecules, such as exosomes, are forced around the porous particles and are eluted from the column first. Ultrafiltration, on the other hand, uses ultrafiltration membranes with different molecular weight cut-offs to selectively separate samples. Despite the simplicity and speed of execution, this method has some limitations, including high membrane clogging and vesicle entrapment that reduce the purity of the eluate.

6.7.4.2.3 Polymer precipitation

The polymer-based isolation system uses hydrophilic polymers that alter the solubility of the exosome, cause its precipitation, and are subsequently isolated by centrifugation or low-speed filtration [352]. Several exosome precipitation kits are currently available using mainly polyethylene glycol. Highly hydrophilic polymers interact with water molecules surrounding exosomes and create a hydrophobic microenvironment, resulting in exosome precipitation. Polymer precipitation is a simple procedure, has a short time frame, and does not require specialized equipment. It is performed in a short time and has a high yield and for these reasons is usable for large volumes [353]. Several limitations, however, are associated with this technique. Among them, it has been highlighted that the use of polymers, which cause the precipitation of other components (nucleic acids, proteins), affects the purity of the sample. To reduce this drawback, pre and postcleaning steps should be performed to improve the preparation.

6.7.4.2.4 Immunoaffinity methods

Immunoaffinity chromatography is a separation and purification technology based on interactions between membrane antigens and immobilized antibodies or membrane receptors and their ligands. Elution conditions, biological affinity, and matrix carriers determine binding efficiency [352]. For best performance, antibodies should be fixed on a solid surface, such as chromatography, microspheres, plates, and various types of microfluidic apparatus [352]. Although immunoaffinity ensures high purity, the high cost of antibodies makes the technique rather expensive. In addition, exosome elution using nonphysiological buffers and nonneutral pH could damage exosomes.

6.7.4.2.5 Microfluidics-based methods

This technique uses the physical and biochemical properties of exosomes, such as size, density, and immunoaffinity, to isolate exosomes using microfluidic devices. Over the past decade, among the various forms of microfluidics, immunoaffinity has been the most commonly used. This method involves the specific recognition of exosome markers by corresponding antibodies immobilized on the chips, thus enabling the isolation of specific exosomes from all other extracellular vesicles. Although highly efficient, inexpensive, and easily automated, the technique has a low sampling capacity.

6.7.5 Standardization initiatives and ISO/CEN/external quality assessment development in liquid biopsy

The standardization procedure in liquid biopsy is highly required to ensure the clinical validity of the diagnostic algorithms, in both preanalytical and analytical phases. In this field, the preanalytical step has been defined as the most crucial and limiting for liquid biopsy applications [354]. This phase has been deeply investigated, but no consensus has been yet reached for workflow optimization. Different international organizations are working on initiatives aimed at standardization in liquid biopsy. More of these programs are ongoing and will improve the share of data and the reach of a worldwide consensus for diagnostic purposes. In the international liquid biopsy scenario is fundamental the role of the Foundation for the National Institutes of Health (FNIH) Biomarkers Consortium (The Latest | The Foundation for the National Institutes of Health (fnih.org) founder of the International Liquid Biopsy Standardization Alliance (ILSA) Collaborative Community (Fig. 6.7.3). The community includes organizations believing that working together and sharing data is the key to the diffusion of liquid biopsy and the identification of reference standards ubiquitously for the medical community. ILSA has received support and direct participation from the European Medicines Agency (EMA), and the Center for Devices and Radiological Health (CDRH) of the Food and Drug Administration (FDA). The major entities bring participants together to achieve a common outcome promoting a consensus development [302].

The landscape of international initiatives allowing an opportunity for standardization on liquid biopsy includes:

- **FNIH Biomarkers Consortium ctDNA Quality Control Materials Project** (https://fnih.org/), is an American initiative working by bringing together private, government, academic, and nonprofit entities to understand which could be the most appropriate quality control materials. This

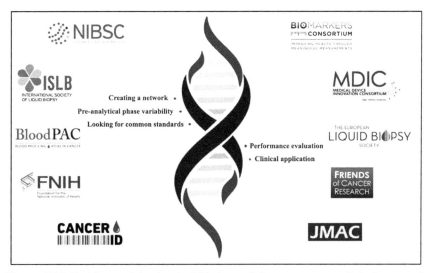

Figure 6.7.3 Entities participating in ILSA that collaborate on preanalytical step variability and other critical issues to standardize liquid biopsy on a global scale by finding common standards and materials for use in precision medicine.

initiative seeks to demonstrate QCM performance as comparable to ctDNA and their suitability for establishing performance characteristics for NGS workflows currently approved and introduced in clinical practice.

- **The Blood Profiling Atlas in Cancer** (**BloodPAC**) BLOODPAC, is an American consortium aimed at accelerating the development, validation, and clinical applications of liquid biopsy workflows. BloodPAC is based on a collaborative network skilled in data sharing among the public, industry, academia, and regulatory entities. This initiative creates the BloodPAC Data Common, a platform useful for contributing and identifying minimum technical data elements in the preanalytical phase and patient status. The interest is to involve liquid biopsy in patient management and outcome improvement by generating more evidence to quickly move the approval process.

- **European Liquid Biopsy Society** (**ELBS**) (UKE—ELBS—Project) has replaced the successful EU/IMI CANCER-ID consortium. CANCER-ID was a European consortium funded by the Innovative Medicines Initiative that wanted to find an alternative strategy for monitoring neoplastic diseases. Specifically, it worked to develop and clinically validate standard blood-based biomarkers workflows, such as procedures based on CTCs, cf/ctDNA, and other biomarkers such as micro-RNAs. In 2020, the ELBS was introduced at the University

Hospital Hamburg-Eppendorf (UKE). The membership derived from academic institutions and industry partners agreeing to promote the study of liquid biopsy and its use in clinical patient management.

- **International Society of Liquid Biopsy** (ISLB) (https://www.isliquidbiopsy.org/) is an organization founded in Granada in 2017 to bring together oncology and other clinical specialties with liquid biopsy tools. An important step of ISLB was the introduction of liquid biopsy recommendations to characterize and develop diagnostics and prognostics instruments improving collective healthcare and patient.

- **Friends of Cancer Research** (Homepage | Friends of Cancer Research) has been an initiative trying to standardize ctDNA tests considering this as indicative of outcomes. Friends of Cancer Research developed the initiative ctDNA to MONItor Treatment Response Project aiming at using ctDNA to monitor patient treatment response.

- **The Japanese bio–Measurement and Analysis Consortium** (http://www.jmac.or.jp) is a different initiative thought as an industrial group supporting and improving biotechnology with international standardization purposes.

- **The Medical Device Innovation Consortium** [Medical Device Innovation Consortium | Regulatory Science Advancement (mdic.org)] was born as a public-private collaboration creating advanced medical device regulatory aspects for liquid biopsy and other medical areas. For example, an important goal was the development of somatic reference samples in solid tissues.

- **NIBSC** (NIBSC—Home) represents the WHO International Laboratory for Biological Standards. The main goal of this initiative is to develop the first International Standard for ctDNA on Biological Standardization approved by the WHO Expert Committee regarding the calibration of the available assays and secondary standards. The primary intention is to produce ctDNA standards for the most common clinically EGFR-associated sequence variants, such as p.L858R, p.T790M, and exon 19 deletions, to achieve a broader landscape extended to align liquid and solid tumor biopsy. NIBS is also focused on the study of matrices, and it is currently evaluating the performance of different arrays with fragmented DNA derived from various cell lines to determine the optimal standards, cross-referenced to patient ctDNA. Standardization involves identifying and achieving harmonization of the multiple variables associated with analysis and measurement, including the percentage of variants, DNA fragment size, ctDNA yield, and gene copy number. Stable storage and comparison with real samples is also a critical issue as the assays must work in all laboratories.

Horizon 2020 SPIDA4P is a 48-month project built on SPIDIA's results continuing the mantle of SPIDIA (2008—2013). It has been funded by the European Union FP7 program and it is focused on standardizing and improving *in vitro* diagnostic preanalytical workflows. The main goal is to enhance the global healthcare system by developing preanalytical standards 22 pan-European documents fundamental for the European Committee for Standardization (CEN) and the International Organization for Standardization (ISO), as important as developing external quality assessment (EQAs) schemes and implementation instruments. SPIDIA4P has involved 19 distinguished academic institutions, organizations with international influence, and life sciences companies, with the coordination offered by QIAGEN GmbH. Thanks to this initiative a lot of new preanalytical technologies have been projected. Regarding the CEN/TC 140 for "In Vitro Medical Devices," SPIDIA has been allowed to develop and introduce CEN technical specifications (CEN/TS) for preanalytical procedures in Europe. Furthermore, SPIDIA4P has seen the realization and release of several ISO standards by the ISO/TC 212 "Clinical Laboratory Testing and *In Vitro* Diagnostic Test Systems" with the EN/ISO standards, along with several EQAs for the evaluation of the performance of specific in vitro diagnostic procedures.

6.7.6 Key points

- The preanalytical phase has been defined as the most crucial and limiting level for liquid biopsy applications;
- The lack of strong evidence of clinical utility and validity and the lack of standardized methods still limits the clinical success of liquid biopsy;
- One of the most common difficulties related to the cfDNA and CTCs is their fragility which leads to their degradation within hours from the collection;
- Exosome characterization and isolation are technically challenging: preanalytical variables can affect exosome size, morphology, and yield;
- The standardization procedure in liquid biopsy is highly required to ensure the clinical validity of the diagnostic algorithms.

6.7.7 Expert opinion

Nowadays, with the advent of precision medicine, above all in the oncology field, liquid biopsy represents a valid and intelligent alternative to solid biopsies, especially at a molecular level. Unfortunately, the lack of strong evidence of clinical utility and validity along with the absence of a recognized consensus in standardization methods and workflow optimization still limits the clinical success of liquid biopsy. Moreover, the standardization procedure in liquid biopsy is highly required to ensure the clinical validity of the diagnostic algorithms, and to obtain a consensus between different laboratories worldwide. Furthermore, the preanalytical phase has been defined as the most crucial and limiting step for liquid biopsy applications, especially for cfDNA, CTCs, and Exosome analysis. To reach greater accuracy and homogeneity in workflows, different international organizations are working on initiatives aimed at standardization in liquid biopsy. In particular, more of these programs are ongoing and will improve the share of data and the reach of a worldwide consensus for diagnostic purposes. In particular, the International Liquid Biopsy Standardization Alliance Collaborative Community includes organizations believing that working together and sharing data is the key to the diffusion of liquid biopsy and the identification of reference standards ubiquitously for the medical community. Moreover, ILSA has received support and direct participation from the EMA, and the CDRH of the FDA. The major entities bring participants together to achieve a common outcome promoting a consensus development.

Acknowledgments

M. La Mantia, C. Brando, A. Fiorino, E. Pedone, and V. Gristina contributed to the current work under the Doctoral Program in Experimental Oncology and Surgery, University of Palermo.

References

[302] Connors D, Allen J, Alvarez JD, Boyle J, Cristofanilli M, Hiller C, et al. International liquid biopsy standardization alliance white paper. Crit Rev Oncol Hematol 2020;156:103112.

[303] Gristina V, Malapelle U, Galvano A, Pisapia P, Pepe F, Rolfo C, et al. The significance of epidermal growth factor receptor uncommon mutations in non-small cell lung cancer: a systematic review and critical appraisal. Cancer Treat Rev 2020;85:101994.

[304] Gristina V, La Mantia M, Galvano A, Cutaia S, Barraco N, Castiglia M, et al. Non-Small cell lung cancer harboring concurrent EGFR genomic alterations: a systematic review and critical appraisal of the double dilemma. J Mol Pathol 2021;2 (2):173−96. Available from: https://www.mdpi.com/2673−5261/2/2/16.

[305] Galvano A, Gristina V, Malapelle U, Pisapia P, Pepe F, Barraco N, et al. The prognostic impact of tumor mutational burden (TMB) in the first-line management of advanced non-oncogene addicted non-small-cell lung cancer (NSCLC): a systematic review and meta-analysis of randomized controlled trials. ESMO Open 2021;6(3). Available from: https://doi.org/10.1016/j.esmoop.2021.100124.

[306] Pisapia P, Pepe F, Gristina V, La Mantia M, Francomano V, Russo G, et al. A narrative review on the implementation of liquid biopsy as a diagnostic tool in thoracic tumors during the COVID-19 pandemic Mediastinum 2021;5. Mediastinum [Internet]. 2021. Available from: https://med.amegroups.com/article/view/6433.

[307] Mantia MLa, Koyyala VPB. The war against coronavirus disease 19 through the eyes of cancer physician: an Italian and indian young medical oncologist's perspective. Indian J Med Paediatr Oncol 2020;41(03):305−7.

[308] Gristina V, La Mantia M, Iacono F, Galvano A, Russo A, Bazan V. The emerging therapeutic landscape of ALK inhibitors in non-small cell lung cancer. Pharmaceuticals (Basel) 2020;13(12).

[309] Russo A, Incorvaia L, Del Re M, Malapelle U, Capoluongo E, Gristina V, et al. The molecular profiling of solid tumors by liquid biopsy: a position paper of the AIOM-SIAPEC-IAP-SIBioC-SIC-SIF Italian Scientific Societies. ESMO open 2021;6(3):100164.

[310] Russo A, Incorvaia L, Malapelle U, Del Re M, Capoluongo E, Vincenzi B, et al. The tumor-agnostic treatment for patients with solid tumors: a position paper on behalf of the AIOM- SIAPEC/IAP-SIBioC-SIF Italian Scientific Societies. Crit Rev Oncol Hematol 2021;165:103436.

[311] Russo A, Incorvaia L, Capoluongo E, Tagliaferri P, Galvano A, Del Re M, et al. The challenge of the molecular tumor board empowerment in clinical oncology practice: a position paper on behalf of the AIOM- SIAPEC/IAP-SIBioC-SIC-SIF-SIGU-SIRM Italian Scientific Societies. Crit Rev Oncol Hematol 2022;169:103567. Available from: https://www.sciencedirect.com/science/article/pii/S1040842821003541.

[312] Beretta G, Capoluongo E, Danesi R, Del Re M, Fassan M, Giuffrè G, et al. Raccomandazioni 2020 per l'esecuzione di Test Molecolari su Biopsia Liquida in Oncologia, 2020.

[313] Pisapia P, Pepe F, Baggi A, Barberis M, Galvano A, Gristina V, et al. Next generation diagnostic algorithm in non-small cell lung cancer predictive molecular pathology: the KWAY Italian multicenter cost evaluation study. Crit Rev Oncol Hematol 2021;169:103525.

[314] Gristina V, Galvano A, Castellana L, Insalaco L, Cusenza S, Graceffa G, et al. Is there any room for PD-1 inhibitors in combination with platinum-based chemotherapy as frontline treatment of extensive-stage small cell lung cancer? A systematic review and *meta*-analysis with indirect comparisons among subgroups and landmark survival a. Ther Adv Med Oncol 2021;13. Available from: https://doi.org/10.1177/17588359211018018 17588359211018018.

[315] Galvano A, Castiglia M, Rizzo S, Silvestris N, Brunetti O, Vaccaro G, et al. Moving the target on the optimal adjuvant strategy for resected pancreatic cancers: a systematic review with *meta*-analysis. Cancers (Basel) 2020;12(3).

[316] Ungerer V, Bronkhorst AJ, Holdenrieder S. Preanalytical variables that affect the outcome of cell-free DNA measurements. Crit Rev Clin Lab Sci 2020;57(7):484−507.

[317] Passiglia F, Rizzo S, Rolfo C, Galvano A, Bronte E, Incorvaia L, et al. Metastatic site location influences the diagnostic accuracy of ctDNA EGFR- mutation testing in nsclc patients: a pooled analysis. Curr Cancer Drug Targets 2018;18(7):697−705.

[318] Bronkhorst AJ, Ungerer V, Holdenrieder S. The emerging role of cell-free DNA as a molecular marker for cancer management. Biomol Detect Quantif 2019;17:100087.

[319] Leto G, Incorvaia L, Flandina C, Ancona C, Fulfaro F, Crescimanno M, et al. Clinical impact of cystatin C/Cathepsin L and follistatin/activin A systems in breast cancer progression: a preliminary report. Cancer Invest 2016;34(9):415−23.

[320] Lampignano R, Neumann MHD, Weber S, Kloten V, Herdean A, Voss T, et al. Multicenter evaluation of circulating cell-free DNA extraction and downstream analyses for the development of standardized (pre)analytical work flows. Clin Chem 2020;66(1):149−60.

[321] Badalamenti G, Barraco N, Incorvaia L, Galvano A, Fanale D, Cabibi D, et al. Are Long noncoding RNAs new potential biomarkers in gastrointestinal stromal tumors (GISTs)? the role of H19 and MALAT1. J Oncol 2019;2019:5458717.

[322] Incorvaia L, Fanale D, Badalamenti G, Porta C, Olive D, De Luca I, et al. Baseline plasma levels of soluble PD-1, PD-L1, and BTN3A1 predict response to nivolumab treatment in patients with metastatic renal cell carcinoma: a step toward a biomarker for therapeutic decisions. Oncoimmunology 2020;9(1):1832348.

[323] Sorber L, Zwaenepoel K, Deschoolmeester V, Van Schil PEY, Van Meerbeeck J, Lardon F, et al. Circulating cell-free nucleic acids and platelets as a liquid biopsy in the provision of personalized therapy for lung cancer patients. Lung Cancer 2017;107:100−7.

[324] Massihnia D, Galvano A, Fanale D, Perez A, Castiglia M, Incorvaia L, et al. Triple negative breast cancer: shedding light onto the role of pi3k/akt/mtor pathway. Oncotarget 2016;7(37):60712−22.

[325] Galvano A, Scaturro D, Badalamenti G, Incorvaia L, Rizzo S, Castellana L, et al. Denosumab for bone health in prostate and breast cancer patients receiving endocrine therapy? A systematic review and a meta-analysis of randomized trials. J Bone Oncol. 2019;18:100252. Available from: https://doi.org/10.1016/j.jbo.2019.100252. PMID: 31440444; PMCID: PMC6700425.

[326] Passiglia F, Calandri M, Guerrera F, Malapelle U, Mangone L, Ramella S, et al. Lung cancer in Italy. J. Thoracic Oncol. 2019;14:2046−52.

[327] Bronte G, Incorvaia L, Rizzo S, Passiglia F, Galvano A, Rizzo F, et al. The resistance related to targeted therapy in malignant pleural mesothelioma: why has not the target been hit yet? Crit Rev Oncol Hematol 2016;107:20−32.

[328] Ignatiadis M, Rack B, Rothé F, Riethdorf S, Decraene C, Bonnefoi H, et al. Liquid biopsy-based clinical research in early breast cancer: the EORTC 90091-10093 Treat CTC trial. Eur J Cancer 2016;63:97−104.

[329] Incorvaia L, Fanale D, Badalamenti G, Barraco N, Bono M, Corsini LR, et al. Programmed death ligand 1 (PD-L1) as a predictive biomarker for pembrolizumab therapy in patients with advanced non-small-cell lung cancer (NSCLC). Adv Ther 2019.

[330] Malapelle U, Mayo de-Las-Casas C, Rocco D, Garzon M, Pisapia P, Jordana-Ariza N, et al. Development of a gene panel for next-generation sequencing of clinically relevant mutations in cell-free DNA from cancer patients. Br J Cancer 2017;116(6):802−10.

[331] Pisapia P, Malapelle U, Troncone G. Liquid biopsy and lung cancer. Acta Cytol 2019;63(6):489−96.

[332] Vigliar E, Iaccarino A, Bruzzese D, Malapelle U, Bellevicine C, Troncone G. Cytology in the time of coronavirus disease (covid-19): an Italian perspective. J Clin Pathol 2020.

[333] Scilla KA, Rolfo C. The role of circulating tumor DNA in lung cancer: mutational analysis, diagnosis, and surveillance now and into the future. Curr Treat Options Oncol 2019;20(7):61.

[334] Rolfo C, Giovannetti E, Hong DS, Bivona T, Raez LE, Bronte G, et al. Novel therapeutic strategies for patients with NSCLC that do not respond to treatment with EGFR inhibitors. Cancer Treat Rev 2014.

[335] Pagani O, Francis PA, Fleming GF, Walley BA, Viale G, Colleoni M, et al. Absolute improvements in freedom from distant recurrence to tailor adjuvant endocrine therapies for premenopausal women: results from TEXT and SOFT. J Clin Oncol J Am Soc Clin Oncol 2020;38(12):1293−303.

[336] Bettegowda C, Sausen M, Leary RJ, Kinde I, Wang Y, Agrawal N, et al. Detection of circulating tumor DNA in early- and late-stage human malignancies. Sci Transl Med 2014.

[337] Fisher B, Jeong J-H, Dignam J, Anderson S, Mamounas E, Wickerham DL, et al. Findings from recent national surgical adjuvant breast and bowel project adjuvant

studies in stage I breast cancer. JNCI Monogr 2001;2001(30):62—6. Available from: https://doi.org/10.1093/oxfordjournals.jncimonographs.a003463.

[338] Ahrendt SA, Hu Y, Buta M, McDermott MP, Benoit N, Yang SC, et al. p53 mutations and survival in stage I non-small-cell lung cancer: results of a prospective study. J Natl Cancer Inst 2003;95(13):961—70.

[339] Rothé F, Laes JF, Lambrechts D, Smeets D, Vincent D, Maetens M, et al. Plasma circulating tumor DNA as an alternative to metastatic biopsies for mutational analysis in breast cancer. Ann Oncol 2014;25(10):1959—65.

[340] Cristofanilli M, Hayes DF, Budd GT, Ellis MJ, Stopeck A, Reuben JM, et al. Circulating tumor cells: a novel prognostic factor for newly diagnosed metastatic breast cancer. J Clin Oncol J Am Soc Clin Oncol 2005;23(7):1420—30.

[341] Rijavec E, Coco S, Genova C, Rossi G, Longo L, Grossi F. Liquid Biopsy in Non-Small cell lung cancer: highlights and challenges. Cancers (Basel) 2019;12(1).

[342] Tinari N, Fanizza C, Romero M, Gambale E, Moscetti L, Vaccaro A, et al. Identification of subgroups of early breast cancer patients at high risk of nonadherence to adjuvant hormone therapy: results of an Italian survey. Clin Breast Cancer 2015;15(2):e131—7.

[343] Malapelle U, Pepe F, Pisapia P, Sgariglia R, Nacchio M, Barberis M, et al. Next generation diagnostic algorithm in non-small cell lung cancer predictive molecular pathology: the KWAY Italian multicenter cost evaluation study. Crit Rev Oncol Hematol 2005;169(7):103525.

[344] Pisapia P, Pepe F, Baggi A, Barberis M, Galvano A, Gristina V, et al. Next generation diagnostic algorithm in non-small cell lung cancer predictive molecular pathology: the KWAY Italian multicenter cost evaluation study. Crit Rev Oncol Hematol 2022;169:103525.

[345] Freitas C, Sousa C, Machado F, Serino M, Santos V, Cruz-Martins N, et al. The Role of liquid biopsy in early diagnosis of lung cancer. Front Oncol 2021;11:1130. Available from: https://www.frontiersin.org/article/10.3389/fonc.2021.634316.

[346] Galvano A, Taverna S, Badalamenti G, Incorvaia L, Castiglia M, Barraco N, et al. Detection of RAS mutations in circulating tumor DNA: a new weapon in an old war against colorectal cancer. A systematic review of literature and *meta*-analysis. Ther Adv Med Oncol 2019;11 1758835919874653.

[347] Maas SLN, Breakefield XO, Weaver AM. Extracellular vesicles: unique intercellular delivery vehicles. Trends Cell Biol 2017;27(3):172—88.

[348] Chen M, Xu Y, Zhao J, Zhong W, Zhang L, Bi Y, et al. Concurrent driver gene mutations as negative predictive factors in epidermal growth factor receptor-positive non-small cell lung cancer. EBioMedicine 2019;42:304—10. Available from: https://www.sciencedirect.com/science/article/pii/S235239641930163X.

[349] Ludwig N, Whiteside TL, Reichert TE. Challenges in exosome isolation and analysis in health and disease. Int J Mol Sci 2019;20(19).

[350] Yang D, Zhang W, Zhang H, Zhang F, Chen L, Ma L, et al. Progress, opportunity, and perspective on exosome isolation—efforts for efficient exosome-based theranostics. Theranostics 2020;10(8):3684—707.

[351] Incorvaia L, Russo A, Cinieri S. The molecular tumor board: a tool for the governance of precision oncology in the real world. Tumori U S 2021. p. 3008916211062266.

[352] Zhang Y, Bi J, Huang J, Tang Y, Du S, Li P. Exosome: a review of its classification, isolation techniques, storage, diagnostic and targeted therapy applications. Int J Nanomed 2020;15:6917—34.

[353] Garon EB, Ciuleanu T, Arrieta O, Prabhash K, Syrigos KN, Goksel T, et al. Ramucirumab plus docetaxel vs placebo plus docetaxel for second-line treatment of stage IV non-small-cell lung cancer after disease progression on platinum-based therapy (REVEL): a multicentre, double-blind, randomised phase 3 trial. Lancet. 2014.

[354] Serrano MJ, Garrido-Navas MC, Diaz Mochon JJ, Cristofanilli M, Gil-Bazo I, Pauwels P, et al. Precision prevention and cancer interception: the new challenges of liquid biopsy. Cancer Discov 2020;10(11):1635−44.

Further reading

International liquid biopsy standardization alliance white paper. https://pubmed.ncbi.nlm.nih.gov/33035734/.

The pre-analytical phase of the liquid biopsy. https://pubmed.ncbi.nlm.nih.gov/31580920/.

CHAPTER 7

Early detection screening: myth or reality?

M. La Mantia[1], F. Iacono[1], S. Cutaia[1], V. Gristina[1], A. Perez[1], M. Greco[1], K. Calcara[1], A. Galvano[1], V. Bazan[2] and A. Russo[1]

[1]Department of Surgical, Oncological, and Oral Sciences, University of Palermo, Palermo, Italy
[2]Department of Biomedicine, Neuroscience and Advanced Diagnostics (Bi.N.D.), University of Palermo, Palermo, Italy

Historically, surgical biopsy has been considered the gold standard for cancer diagnosis. Nonetheless, surgical biopsies are not always feasible, and the procedures are often expensive, sometimes leading to clinical complications for the patient. Moreover, a patient's tumor genomic profile dynamically evolves, especially under oncological treatments, delineating the hurdle of tumor heterogeneity. In this fascinating scenario, liquid biopsy has emerged as a novel and intriguing diagnostic tool, providing a significant step forward in the field due to lower cost, accessibility, and repeatability. Over the last decades, several essential trials have affirmed the efficacy and the importance of the radiological screening by low dose computed tomographic for the lung cancer early detection, demonstrating to reduce the mortality among subjects at high risk of disease [1]. In fact, the US-based National Lung Screening Trial reported that three annual computed tomographic (CT) screening exams produced 20.0% lower mortality from lung cancer than screening with the use of chest radiography alone among 53,454 participants at high risk for lung cancer after a median follow-up of 6.5 years. The study recently assessed that mortality after a follow-up of 5.5 and 6.0 years was close to 19% lower with CT than with chest radiography [2,3]. Moreover, similar experiences have also been reported in Europe, particularly in The Dutch−Belgian lung-cancer screening NELSON trial, a population-based, randomized, controlled trial initiated in 2000, aimed to show shrinkage in lung-cancer mortality of 25% or more with volume-based, low-dose CT lung cancer screening in high-risk male patients after ten years of follow-up [4−6].

Liquid Biopsy
DOI: https://doi.org/10.1016/B978-0-12-822703-9.00001-6

© 2023 Elsevier Inc.
All rights reserved.

However, different concerns affect the routinary use of radiological screening tools, such as the high-false positive rate, the potential for over-diagnosis, costs, and the radiation exposure field [7,8].

Therefore, it is required to implement the lung cancer screening scenario with less invasive, sensitive, and specific diagnostic tools such as detecting biomarkers. In this context, the usefulness of liquid biopsy in consistently attaining utmost attention when it comes to timely detection and diagnosis of non-small cell lung carcinoma (NSCLC). Furthermore, blood specimen analysis or liquid biopsy has extensively been speculated as a matter of early and convenient NSCLC screening worldwide. Some evidence supported that CTC may be a potential screening tool to detect lung cancer in the early stage [9]. Ilie et al. studied CTC assessment in patients with chronic obstructive pulmonary disease. 3% of patients were detected with CTC, and all of these patients developed lung cancer within four years. On the contrary, none of the CTC-negative patients developed cancer within the monitored period [10−12].

The cfDNA has a pretty controversial role in the landscape of screening for Lung cancer compared with other liquid biopsy components due to some limitations: cfDNA is markedly diluted compared to germline circulating DNA, is frequently present in the circulation in healthy individuals, thus in smaller concentrations, it positively correlates with tumor size and staging and, as well, cfDNA levels are increased in some benign or premalignant conditions included other lung diseases. CfDNA represents tumor heterogeneity and dynamics; genetic and epigenetic alterations reflect the original tumor [13,14]. However, despite the use of assay with perfect analytical sensitivity when the level of ctDNA is too low, the detection of mutant loci could lead to a negative result. It has been demonstrated that miRNAs are valid biomarkers in discrimination between healthy individuals and lung cancer patients, with good potential as predictor factors of diagnosis and prognosis [15,16].

A study by Shen et al. recruited 65 individuals, who were selected based on the presence of solitary pulmonary nodules (SPNs) on chest CT scans. Final diagnoses were confirmed with histopathologic examinations of specimens. The study demonstrated that circulating miR-486-5p expression was significantly reduced in the blood plasma of malignant SPNs patients compared to subjects with benign SPNs and healthy controls. These three circulating miRNAs demonstrated 75% specificity and 84.95% accuracy in differentiating LC patients from healthy individuals [17−19].

Of note, some results from a cohort of a recent retrospective study recently published on Jama Oncology showed that the signature of 14 miRNAs had identified patients with early-stage lung cancer with an accuracy of 95.9% (95% CI, 95.7%−96.2%), a sensitivity of 76.3% (95% CI, 74.5%−78.0%), and a specificity of 97.5% (95% CI, 97.2%−97.7%) [20,21]. Equally, exosomes are valid candidates in lung cancer early diagnosis, mainly because of their stability and accessibility in most types of body fluids [15,22].

Liquid biopsy has an emerging role in detecting any tumor-derived material in the blood after curative treatment. Minimal residual disease (MRD) is a potential and powerful tool for selecting patients for adjuvant therapy to detect earlier the relapse of disease and consequently anticipate the start of treatment, and tailor adjuvant therapies, avoiding the adjuvant chemotherapy in low-risk patients and administrating the therapy only to those at high-risk [23,24].

ctDNA is a good candidate for these purposes. It has been demonstrated that the decrease in ctDNA in patients undergoing tumor resection and the detection of an increase during the follow-up could indicate an evolving relapse of disease [25,26]. There is a demonstrable decrease in ctDNA in patients undergoing tumor resection for NSCLC, and monitoring for an increase could indicate an evolving relapse [27,28].

In one prospective study, postsurgical monitoring of ctDNA following curative primary resection resulted in over half of patients having detectable ctDNA posttreatment. On the contrary, 72% of patients had detectable ctDNA before relapse. Additionally, the detection of ctDNA predated radiological evidence of deterioration by a median of 5.3 months [29,30].

Chen et al. conducted a small prospective study in which 76 pts with suspected lung cancer who underwent curative-intent lung resection were prospectively enrolled. Targeted DNA sequencing with a next-generation sequencing platform identified a series of somatic mutations in matched tumor tissue DNA (ctDNA) and plasma ctDNA samples. Plasma was collected at different time points (prior, during, and postsurgery). Among all patients with lung cancer included, 31 had concordant mutations, and 21 had no modification in ctDNA and ctDNA, reaching an overall concordance of 68.4% [27,31].

These data had confirmed. Overall, the results reported by Guo and colleagues who used similar sequencing methods in 41 patients with lung cancer undergoing surgery and showed a concordance of 78%, specificity of 69%, and sensitivity of 93% [32,33].

Abbosh et al. conducted a phylogenetic approach to ctDNA profiling in early-stage NSCLC in the first 100 TRACERx study participants using multiregion exome sequencing (M-Seq).

This work confirmed that plasma genotyping could detect MRD/recurrence weeks to months before imaging. Therefore, the abovementioned clinical trial showed that necrosis, lymphovascular invasion, Ki67 labeling index, tumor size, and nonadenocarcinoma status might anticipate ctDNA drop. Lastly, this assay detected the phylogenetic subclone responsible for the recurrence of the disease. Noteworthy, in the study, tumor cfDNA was placed just in 48% of patients with early-stage NSCLC, and the authors discussed a theoretical limit of DNA shed for tumors of a given size, estimating that tumors <10 cm^3 will not have detectable ctDNA [34,35].

Although many open, challenging issues remain, several advances in methods led to a meaningful improvement of the sensibility and specificity of ctDNA detection and identification. Such us as the optimal timing of ctDNA testing, the definition of a "positive" MRD and the establishment of a unique plasma level ctDNA cut-off which is critical to minimize variations among studies, and finally, the interpretation of a negative ctDNA considering the chance of false-negative test [36,37] (Table 7.1).

Table 7.1 Overview of the ongoing clinical trials on liquid biopsy in the early setting.

Study title	Conditions	Status	ID
Accelerating lung cancer diagnosis through liquid biopsy	NSCLC	Recruiting	NCT04863924
Assessment of early-detection based on liquid biopsy in lung cancer (ASCEND-LUNG)	NSCLC	Recruiting	NCT04817046
Surveillance with PET/CT and liquid biopsies of stage I—III lung cancer patients after completion of definitive therapy	NSCLC	Recruiting	NCT03740126
A preliminary study on the detection of plasma markers in early diagnosis for lung cancer	NSCLC	Recruiting	NCT04558255
The tracking molecular evolution for NSCLC (T-MENC) study	NSCLC	Recruiting	NCT03838588
Fluid biopsy for the diagnosis of lung cancer	NSCLC	Recruiting	NCT04162678
Liquid biopsy with PET/CT vs PET/CT alone in diagnosis of small lung nodules	NSCLC	Recruiting	NCT05066776
ORACLE: Observation of ResiduAl Cancer with Liquid Biopsy Evaluation	Urothelial carcinoma NSCLC Endometrial carcinoma Renal cell carcinoma	Recruiting	NCT05059444

CT, Computed tomography; *NSCLC*, Non-small cell lung cancer; *PET*, Positron Emission Tomography.

References

[1] Russo A, Incorvaia L, Capoluongo E, Tagliaferri P, Galvano A, Del Re M, et al. The challenge of the molecular tumor board empowerment in clinical oncology practice: a position paper on behalf of the AIOM- SIAPEC/IAP-SIBioC-SIC-SIF-SIGU-SIRM Italian scientific societies. Crit Rev Oncol Hematol [Internet] 2022;169:103567. Available from: https://www.sciencedirect.com/science/article/pii/S1040842821003541.

[2] Reduced lung-cancer mortality with low-dose computed tomographic screening. N. Engl. J. Med. 2011.

[3] Passiglia F, Galvano A, Rizzo S, Incorvaia L, Listì A, Bazan V, et al. Looking for the best immune-checkpoint inhibitor in pre-treated NSCLC patients: an indirect comparison between nivolumab, pembrolizumab and atezolizumab. Int J Cancer 2018.

[4] Ru Zhao Y, Xie X, de Koning HJ, Mali WP, Vliegenthart R, Oudkerk M. NELSON lung cancer screening study. Cancer Imaging Publ Int Cancer Imaging Soc 2011;11(1A):S79–84 Spec No.

[5] Incorvaia L, Fanale D, Bono M, Calò V, Fiorino A, Brando C, et al. BRCA1/2 pathogenic variants in triple-negative vs luminal-like breast cancers: genotype-phenotype correlation in a cohort of 531 patients. Ther Adv Med Oncol 2020;12 1758835920975326.

[6] Gristina V, La Mantia M, Iacono F, Galvano A, Russo A, Bazan V. The emerging therapeutic landscape of ALK inhibitors in non-small cell lung cancer. Pharm (Basel) 2020;13(12).

[7] Aberle DR, Abtin F, Brown K. Computed tomography screening for lung cancer: has it finally arrived? Implications of the national lung screening trial. J Clin Oncol J Am Soc Clin Oncol 2013;31(8):1002–8.

[8] Pisapia P, Pepe F, Gristina V, La Mantia M, Francomano V, Russo G, et al. A narrative review on the implementation of liquid biopsy as a diagnostic tool in thoracic tumors during the COVID-19 pandemic. Mediastinum 5; September 2021. Available from: https://med.amegroups.com/article/view/6433.

[9] Passiglia F, Galvano A, Gristina V, Barraco N, Castiglia M, Perez A, et al. Is there any place for PD-1/CTLA-4 inhibitors combination in the first-line treatment of advanced NSCLC?-a trial-level meta-analysis in PD-L1 selected subgroups. Transl Lung Cancer Res 2021;10(7):3106–19.

[10] Ilie M, Hofman V, Long-Mira E, Selva E, Vignaud J-M, Padovani B, et al. Sentinel' circulating tumor cells allow early diagnosis of lung cancer in patients with chronic obstructive pulmonary disease. PLoS One 2014;9(10):e111597.

[11] Lowe AC. Circulating tumor cells: applications in cytopathology. Surg Pathol Clin 2018;11(3):679–86.

[12] Gristina V, La Mantia M, Galvano A, Cutaia S, Barraco N, Castiglia M, et al. Non-small cell lung cancer harboring concurrent EGFR genomic alterations: a systematic review and critical appraisal of the double dilemma. J Mol Pathol [Internet] 2021; 2(2):173–96. Available from: https://www.mdpi.com/2673-5261/2/2/16.

[13] Freitas C, Sousa C, Machado F, Serino M, Santos V, Cruz-Martins N, et al. The role of liquid biopsy in early diagnosis of lung cancer. Front Oncol [Internet] 2021;11:1130. Available from: https://www.frontiersin.org/article/10.3389/fonc.2021.634316.

[14] La Mantia M, Koyyala VPB. The war against coronavirus disease 19 through the eyes of cancer physician: an Italian and Indian young medical oncologist's perspective. Indian J Med Paediatr Oncol 2020;41(03):305–7.

[15] Rijavec E, Coco S, Genova C, Rossi G, Longo L, Grossi F. Liquid biopsy in non-small cell lung cancer: highlights and challenges. Cancers (Basel) 2019;12(1).

[16] Galvano A, Peri M, Guarini AA, Castiglia M, Grassadonia A, De Tursi M, et al. Analysis of systemic inflammatory biomarkers in neuroendocrine carcinomas of the

lung: prognostic and predictive significance of NLR, LDH, ALI, and LIPI score. Ther Adv Med Oncol 2020;12 1758835920942378.

[17] Shen J, Liu Z, Todd NW, Zhang H, Liao J, Yu L, et al. Diagnosis of lung cancer in individuals with solitary pulmonary nodules by plasma microRNA biomarkers. BMC Cancer [Internet] 2011;11(1):374. Available from: https://doi.org/10.1186/1471-2407-11-374.

[18] Galvano A, Gristina V, Malapelle U, Pisapia P, Pepe F, Barraco N, et al. The prognostic impact of tumor mutational burden (TMB) in the first-line management of advanced non-oncogene addicted non-small-cell lung cancer (NSCLC): a systematic review and *meta*-analysis of randomized controlled trials. ESMO Open [Internet] 2021;6(3). Available from: https://doi.org/10.1016/j.esmoop.2021.100124.

[19] Russo A, Incorvaia L, Malapelle U, Del Re M, Capoluongo E, Vincenzi B, et al. The tumor-agnostic treatment for patients with solid tumors: a position paper on behalf of the AIOM- SIAPEC/IAP-SIBioC-SIF Italian Scientific Societies. Crit Rev Oncol Hematol 2021;165:103436.

[20] Fehlmann T, Kahraman M, Ludwig N, Backes C, Galata V, Keller V, et al. Evaluating the use of circulating microRNA profiles for lung cancer detection in symptomatic patients. JAMA Oncol 2020;6(5):714−23.

[21] Incorvaia L, Madonia G, Corsini LR, Cucinella A, Brando C, Gagliardo C, et al. Challenges and advances for the treatment of renal cancer patients with brain metastases: from immunological background to upcoming clinical evidence on immune-checkpoint inhibitors. Crit Rev Oncol Hematol 2021;163:103390.

[22] Galvano A, Castiglia M, Rizzo S, Silvestris N, Brunetti O, Vaccaro G, et al. Moving the target on the optimal adjuvant strategy for resected pancreatic cancers: a systematic review with *meta*-analysis. Cancers (Basel) 2020;12(3).

[23] Chae YK, Oh MS. Detection of minimal residual disease using ctDNA in lung cancer: current evidence and future directions. J Thorac Oncol Publ Int Assoc Study Lung Cancer 2019;14(1):16−24.

[24] Incorvaia L, Fanale D, Badalamenti G, Porta C, Olive D, De Luca I, et al. Baseline plasma levels of soluble PD-1, PD-L1, and BTN3A1 predict response to nivolumab treatment in patients with metastatic renal cell carcinoma: a step toward a biomarker for therapeutic decisions. Oncoimmunology 2020;9(1):1832348.

[25] Di Capua D, Bracken-Clarke D, Ronan K, Baird A-M, Finn S. The liquid biopsy for lung cancer: state of the art, limitations and future developments. Cancers (Basel) [Internet] 2021;(16):13. Available from: https://www.mdpi.com/2072-6694/13/16/3923.

[26] Incorvaia L, Fanale D, Badalamenti G, Barraco N, Bono M, Corsini LR, et al. Programmed death ligand 1 (PD-L1) as a predictive biomarker for pembrolizumab therapy in patients with advanced non-small-cell lung cancer (NSCLC). Adv Ther 2019.

[27] Badalamenti G, Barraco N, Incorvaia L, Galvano A, Fanale D, Cabibi D, et al. Are long noncoding RNAs new potential biomarkers in gastrointestinal stromal tumors (GISTs)? The role of H19 and MALAT1. J Oncol 2019;2019:5458717.

[28] Pisapia P, Pepe F, Baggi A, Barberis M, Galvano A, Gristina V, et al. Next generation diagnostic algorithm in non-small cell lung cancer predictive molecular pathology: the KWAY Italian multicenter cost evaluation study. Crit Rev Oncol Hematol 2021;169:103525.

[29] Chaudhuri AA, Chabon JJ, Lovejoy AF, Newman AM, Stehr H, Azad TD, et al. Early detection of molecular residual disease in localized lung cancer by circulating tumor DNA profiling. Cancer Discov 2017;7(12):1394−403.

[30] Passiglia F, Bronte G, Bazan V, Natoli C, Rizzo S, Galvano A, et al. PD-L1 expression as predictive biomarker in patients with NSCLC: a pooled analysis. Oncotarget 2016;7(15):19738−47.

[31] Chen K, Zhang J, Guan T, Yang F, Lou F, Chen W, et al. Comparison of plasma to tissue DNA mutations in surgical patients with non-small cell lung cancer. J Thorac Cardiovasc Surg 2017;154(3):1123−31 e2.

[32] Guo N, Lou F, Ma Y, Li J, Yang B, Chen W, et al. Circulating tumor DNA detection in lung cancer patients before and after surgery. Sci Rep [Internet] 2016; 6(1):33519. Available from: https://doi.org/10.1038/srep33519.

[33] Passiglia F, Rizzo S, Rolfo C, Galvano A, Bronte E, Incorvaia L, et al. Metastatic site location influences the diagnostic accuracy of ctDNA EGFR- mutation testing in NSCLC patients: a pooled analysis. Curr Cancer Drug Targets 2018;18(7):697−705.

[34] Abbosh C, Birkbak NJ, Swanton C. Early stage NSCLC—challenges to implementing ctDNA-based screening and MRD detection. Nat Rev Clin Oncol 2018; 15(9):577−86.

[35] Leto G, Incorvaia L, Flandina C, Ancona C, Fulfaro F, Crescimanno M, et al. Clinical impact of cystatin C/cathepsin L and follistatin/activin a systems in breast cancer progression: a preliminary report. Cancer Invest 2016;34(9):415−23.

[36] Guibert N, Pradines A, Favre G, Mazieres J. Current and future applications of liquid biopsy in nonsmall cell lung cancer from early to advanced stages. Eur Respir Rev J Eur Respir Soc 2020;29(155).

[37] Russo A, Incorvaia L, Del Re M, Malapelle U, Capoluongo E, Gristina V, et al. The molecular profiling of solid tumors by liquid biopsy: a position paper of the AIOM-SIAPEC-IAP-SIBioC-SIC-SIF Italian Scientific Societies. ESMO Open 2021;6(3):100164.

CHAPTER 8

Molecular tumor board

M. La Mantia[1,*], G. Busuito[1,*], V. Spinnato[1,*], V. Gristina[1], A. Galvano[1], S. Cutaia[1], N. Barraco[1], A. Perez[1], S. Cusenza[1], L. Incorvaia[1], G. Badalamenti[1], A. Russo[1] and V. Bazan[2]

[1]Department of Surgical, Oncological, and Oral Sciences, University of Palermo, Palermo, Italy
[2]Department of Biomedicine, Neuroscience and Advanced Diagnostics (Bi.N.D.), University of Palermo, Palermo, Italy

Learning objectives

By the end of the chapter the reader will:

- Have learned what a tumor board is and what is its role in modern oncology;
- Known which figures should be part of a tumor board;
- Understand the limitations that tumor board and molecular tumor board may meet in specific settings and how these may be overcome;

Historically, the oncologist together with the surgeon and the radiotherapist played a practically exclusive role in the study and treatment of tumors.

However, the work of the surgeon and radiotherapist was dedicated only to surgical act and radiotherapy treatment of the disease, entrusting the oncologist with the difficult task of facing neoplastic pathology in its complexity: from primary antineoplastic therapy to psychological support.

In recent decades neoplastic pathology has become increasingly complex both on a clinical and molecular level.

Cancer is now considered a dynamic process resulting from a succession of phases involving numerous molecular events and different etiologies (genetic and environmental). Its onset, progression, and prognosis entail a high heterogeneity of causes that, together with the high variability between patients with the same type of tumor and individual therapeutic responses, prevent reliable predictability and limit the therapeutic and preventive abilities of the oncologist.

Thanks to the growing knowledge in biology, biotechnology, and genetics, the laboratory activity has contributed to the identification of new knowledge on cancer cells, such as the possibility of personalizing

---

* La Mantia M, Busuito G, Spinnato V. should be considered equally co-first authors.

Liquid Biopsy
DOI: https://doi.org/10.1016/B978-0-12-822703-9.00006-5

© 2023 Elsevier Inc.
All rights reserved.

cancer treatment, molecular therapy, and target therapy; becoming an essential element in therapeutic choice and prognosis. These new technologies imposed a cultural and operational "revolution" in the strategy of fighting cancer.

The most important aspect of this change is the introduction of a new model of the therapeutic relationship, which involves not only the oncologist but multiple professional figures, each with their own experience, actively required in the decision-making process of an adequate diagnostic and therapeutic workup [1].

This new multidisciplinary approach, called "Tumor board" (MDTs), arises from a collaboration of specialists and subspecialists, from a variety of fields, who meet to review and discuss cancer cases (general or site-specific), in order to provide the best treatment planning and reach a possibly better outcome for each individual cancer patient [2,3] (Fig. 8.1).

TBs represent the involvement of all relevant disciplines in the treatment evaluation and planning process, allowing physicians to improve their approach to specific cancers by reviewing their cases and reconsidering changes in diagnosis and management, sharing them with all board members, helping to ensure judicious use of health care resources [3−6].

The MDT was born in the UK in the 1990s, strengthening in 1995 with the publication of the Calman—Hine plan, which radically reformed the UK's cancer services to concentrate cancer care in the hands of multidisciplinary teams, to ensure a high and uniform standard of care, regardless of where they may live [7−10].

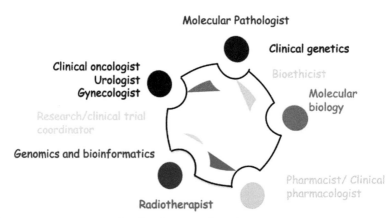

Figure 8.1 Components of molecular tumor board.

Nowadays, telematics technology put at the service of health professionals, as well as some oncological and educational societies (e.g. the European School of Oncology), allows them to meet remotely via videoconference, exchange reports from different locations, to guarantee a greater speed in the coordination of diagnostic choices, also regarding patients who reside in geographically different areas [11−15].

As a consequence, multidisciplinary tumor interventions represent the necessary integrated approach to managing any cancer patient, becoming part of the standard management of cancer care internationally [5,16−19].

Multidisciplinary meetings of the oncological team can include a mix of different specialists cooperating in the same setting, such as thoracic surgeons, medical oncologists, radiotherapists, radiologists, pulmonologists, pathologists, molecular biologists, nuclear medicine specialists, palliative care and rehabilitation doctors [20−23].

As medical practices continue to become more and more specialized, with more complicated treatment options, it becomes necessary for multidisciplinary teams to be as efficient as possible. To achieve this, TBs must provide recommendations that represent the best evidence-based medicine available in the field, even implementing this, if necessary, as a form of the second opinion formulated by expert opinions, when evidence is not available from clinical trials (Table 8.1) [24−26].

For years, the clinical therapeutic benefits of TBs have not been understood, reducing the concept of meeting to something purely theoretical, rather than recognizing an improved and cutting-edge management of patients with cancer, as a result of meetings of sharing, elaboration, and confirmation of data, by specialized figures, each with different skills, animated by a common purpose to guarantee the best diagnostic and therapeutic pathway for patients with malignancy [26,27].

The best cancer institutes cemented the idea of a cancer council and made it mandatory, every patient has to be discussed in the cancer council before treatment is finalized. They have formed management groups for each particular type of cancer.

Multidisciplinary coordination of care is recommended in guidelines from national organizations, including the National Comprehensive Cancer Network [20,21].

TBs have been accredited as part of the fight against cancer management by the American College of Surgeon Cancer Commission as multidisciplinary cancer conferences that prospectively review cases and discuss

Table 8.1 Studies that demonstrated improved outcomes with multidisciplinary management.

Study	Type of study	Endpoints assessed	Outcome 1
Birchall et al. [14]	Retrospective review comparing 2-year survival in H\N cancer pts in the south and west of England before and after a standards document publication (1996–2000)	The 2-year survival for H\N cancer pts	Pts assessed in an MD clinic (consultant oncologist, radiotherapist, and head and neck surgeon) exhibited improved 2-year survival ($P = .03$)
Chang et al. [15]	A retrospective study comparing treatment recommendations before and after an MD breast cancer assessment in 75 consecutive pts	Treatment recommendations made before and after an MD breast cancer assessment	Treatment change in 43% of cases (breast conservation 41% re-excision 6%, further workup 31%, treatment change based pathology review 9%, post-mastectomy RT 9%, HT3%)
Junor et al. [16]	Retrospective population-based analysis of pts with ovarian cancer	Survival of patients with ovarian cancer based on patient factors and organizational/delivery of care factors	Referral to an MD clinic $P < .001$ Receipt of platinum chemotherapy
Lutterbach et al. [13]	A retrospective review of 1516 pts with a brain lesion discussed at an MDB	Assess if recommendations made at Brain TB were implemented	91% of MDB recommendations implemented
Levine et al. [24]	Prospective study of CRC pts (2008–09), comparing pts referred to the MDC versus pts managed outside	The comprehensiveness of the preoperative evaluation, and access to multimodal care	Complete preoperative evaluation in MDC pts was 85% versus 23% in the CG ($P < .0001$)62.5% of MDC pts versus 41.5% of CG pts had peri-operative treatment ($P = .02$) 76% of MDC rectal cancer pts versus 20% of CG pts underwent neoadjuvant therapy ($P < .0001$)

CG, Control group; HT, hormone therapy; MD, multidisciplinary; MDB, multidisciplinary board; MDC, multidisciplinary clinic; RT, radiotherapy; TB, tumor board.

management decisions, intended to encourage participation and improve health outcomes.

Many studies suggest that TBs improve adherence to guidelines by physicians, becoming a reference for discussion and decision making, even considering multiple and up-to-date treatment methods available [5,19−22].

Multidisciplinary groups of experts, therefore, develop high-quality guidelines in cancer care, based on scientific evidence, providing the rationale for specific recommendations (Fig. 8.2). Their spread has made them easily usable in the healthcare delivery system, providing a source of education not only to professionals but also to patients and families. In this way, some studies have shown better treatment results and a decrease in resource utilization resulting in a decrease in the cost of care.

Moreover, the Breast Health Global Initiative in the treatment to be carried out on individual patients with malignancy has developed guidelines differentiated into levels: basic, intermediate, and advanced.

This choice arises from the awareness that elaborating a single advanced guideline would entail implementation problems on an international scale because the financial resources of various health systems are not uniform on a global level. The differentiation in guidelines allows for a worldwide application, so each health system will adopt the guideline based on available resources [28−31].

Another critical aspect of multidisciplinary commissions is the difficulty associated with the concrete execution of a team of specialized professionals able to communicate and collaborate promptly, even remotely, on the treatment to be implemented. On the other hand, metropolitan hospitals are probably able to rise this system as opposed to minor hospitals which, in addition to having limits on economic resources, have evident structural limits such as not allowing advanced coordination [32].

In conclusion, despite the critical issues described above, Tumor Boards represent innovative oncological realities capable of throwing new light on the diagnosis and treatment of cancer through interdisciplinary decisions, they are the best choice for adequate patient care and management plan.

Recent scientific studies have shown that the elaboration process of care plans entrusted to a group of experts, guarantees individual patients an increasingly personalized treatment, suitable for achieving the best final result, as well as better management of available resources, avoiding their waste [33].

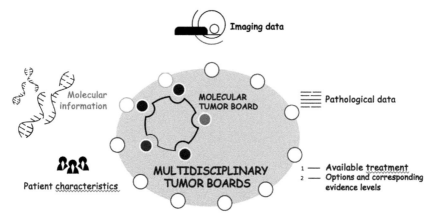

Figure 8.2 Ideal workflow of molecular tumor board.

Once the need for a multidisciplinary oncology team for optimal disease control is established, it is hoped that hospitals, as well as the training of individual professionals and the distribution of resources, will be globally adapted, to ensure uniformity and effectiveness of care regardless of where a patient lives.

Expert opinion

Tumor Boards are an irreplaceable tool to handle challenging oncological cases through confrontation and discussion between different healthcare providers and should be implemented whenever possible.

Molecular tumor boards (MTB) are increasingly needed to guide the treatment selection based on genetic information obtained through high-throughput platforms. MTB should be composed at least of oncologists; hematologists; molecular pathologists; molecular biologists; clinical biologists; geneticists; surgeons; bioinformaticians. Other figures may be included depending on the specific MTB and case discussed. More studies and international guidelines are needed to confirm the available results on the efficacy of MTB and to guide their development worldwide. Data sharing and standardization of procedures, techniques, and workflows are needed to fully take advantage of MTB.

Key points

- Tumor boards derive from the increasing need for specialized expertise and reciprocal cooperation between cancer care providers. It consists of a group of experts of different healthcare disciplines meeting periodically to discuss challenging cases and contribute to the diagnostic and treatment decisions.

- The advances in genomics and the advent of new technologies have vastly increased our knowledge of cancer genetics, however, clinicians have a hard time keeping pace with these developments. MTB aims to fill the gap between genetic data and everyday clinical practice.
- Members can vary depending on the type of board and the specific cases discussed but should be at least composed of an oncologist, surgeon, radiologist, radiation therapy oncologist, or pathologist.
- TB and MTB have been shown to improve "patients" prognosis and outcome measures such as overall survival and progression-free survival. Solid data on MTB are still scarce and there are no international guidelines.

Acknowledgments

M. La Mantia, S. Cutaia, and V. Gristina contributed to the current work under the Doctoral Programme in Experimental Oncology and Surgery, University of Palermo.

References

[1] Keating NL, Landrum MB, Lamont EB, Bozeman SR, Shulman LN, McNeil BJ. Tumor boards and the quality of cancer care.

[2] National Cancer Institute. Definition of tumor board review. http://www.cancer.gov/dictionary?cdrid = 322893 [accessed 01.06.12].

[3] Henson DE, Frelick RW, Ford LG, et al. Results of a national survey of characteristics of hospital tumor conferences. Surg Gynecol Obstet 1990;170(1):1−6.

[4] Calman−Hine Report. A report by the expert advisory group on cancer to the chief medical officers of England and Wales. A policy framework for commissioning cancer services − The Calman−Hine Report. London: Department of Health; 1995.

[5] Morris E, Haward RA, Gilthorpe MS, Craigs C, Forman D. The impact of the Calman−Hine report on the processes and outcomes of care for Yorkshire's colorectal cancer patients. Br J Cancer 2006;95(8):979−85.

[6] Impact of GI Tumor Board on Patient Management and Adherence to Guidelines Haneen A. AlFarhan, Ghada F. Algwaiz, Hajer A. Alzahrani, Roaa S. Alsuhaibani, Ashwaq Alolayan, Nafisa Abdelhafiz, Yosra Ali, Sami Boghdadly, and Abdul Rahman Jazieh.

[7] Tumor Boards: Optimizing the Structure and Improving Efficiency of Multidisciplinary Management of Patients with Cancer Worldwide Nagi S. El Saghir, MD, FACP, Nancy L. Keating, MD, MPH, Robert W. Carlson, MD, Katia E. Khoury, MD, Lesley Fallowfield, DPhil, BSc.

[8] National Cancer Institute. Definition of tumor board review. http://www.cancer.gov/dictionary?cdrid = 322893.

[9] Chang JH, Vines E, Bertsch H, et al. The impact of a multidisciplinary breast cancer center on recommendations for patient management: the University of Pennsylvania experience. Cancer. 2001;91:1231−7.

[10] Wright FC, De Vito C, Langer B, et al. Multidisciplinary cancer conferences: a systematic review and development of practice standards. Eur J Cancer 2007;43:1002−10.

[11] Commission on Cancer. Cancer program standards 2012, version 1.2.1: ensuring patient-centered care. Chicago, IL: American College of Surgeons; 2012.

[12] Jackson GL, Zullig LL, Zafar SY, et al. Using NCCN clinical practice guidelines in oncology to measure the quality of colorectal cancer care in the veterans health administration. J Natl Cpmpr Canc Netw 2013;11:431−41.

[13] Boland GM, Chang GJ, Haynes AB, et al. Association between adherence to National Comprehensive Cancer Network treatment guidelines and improved survival in patients with colon cancer. Cancer 2013;119:1593−601.

[14] Visser BC, Ma Y, Zak Y, et al. Failure to comply with NCCN guidelines for the management of pancreatic cancer compromises outcomes. HPB (Oxford) 2012;14:539−47.

[15] Neubauer MA, Hoverman JR, Kolodziej M, et al. Cost effectiveness of evidence-based treatment guidelines for the treatment of non-small-cell lung cancer in the community setting. J Oncol Pract 2010;6:12−18.

[16] Hoverman JR, Cartwright TH, Patt DA, et al. Pathways, outcomes, and costs in colon cancer: retrospective evaluations in two distinct databases. J Oncol Pract 2011;7:52s−9s.

[17] Newman EA, Guest AB, Helvie MA, et al. Changes in surgical management resulting from case review at a breast cancer multidisciplinary tumor board. Cancer 2006;107:2346−51.

[18] Mulcahy N. Are tumor boards a waste of time? Medscape. http://www.medscape.com/viewarticle/776833; 2012 [accessed 07.01.14].

[19] Lesslie M, Parikh JR. Implementing a multidisciplinary tumor board in the community practice setting. Diagnostics (Basel) 2017;7(4):55−8.

[20] Jazieh AR, Azim HA, McClure J, et al. The Abu Dhabi Declaration: adapted application of NCCN clinical practice guidelines in oncology in the middle east and north Africa Region. http://www.nccn.org/JNCCN/PDF/2010_Vol8_Suppl3_MENA.pdf; [accessed 11.01.14].

[21] National Comprehensive Cancer Network. NCCN clinical practice guidelines in oncology, Asia consensus statement, breast cancer, vol. 1. 2009. http://www.nccn.org/professionals/physician_gls/PDF/breast-asia.pdf; [accessed 15.01.14].

[22] Fallowfield L, Langridge C, Jenkins V. Communication skills training for breast cancer teams talking about trials. Breast 2013;23:193−7.

[23] Jenkins VA, Farewell D, Farewell V, et al. Teams talking trials: results of an RCT to improve the communication of cancer teams about treatment trials. Contemp Clin Trials 2013;35:43−51.

[24] Anderson BO, Yip CH, Smith RA, et al. Guideline implementation for breast healthcare in low-income and middle-income countries: overview of the Breast Health Global Initiative Global Summit 2007. Cancer 2008;113:2221−43.

[25] Eniu A, Carlson RW, El Saghir NS, et al. Guideline implementation for breast healthcare in low- and middle-income countries: treatment resource allocation. Cancer 2008;113:2269−81.

[26] Anderson BO, Cazap E, El Saghir NS, et al. Optimisation of breast cancer management in low-resource and middle-resource countries: executive summary of the Breast Health Global Initiative consensus, 2010. Lancet Oncol 2011;12:387−98.

[27] Fleissig A, Jenkins V, Catt S, et al. Multidisciplinary teams in cancer care: are they effective in the UK? Lancet Oncol 2006;7:935−43.

[28] Russo A, Incorvaia L, Capoluongo E, Tagliaferri P, Galvano A, Del Re M, et al. The challenge of the Molecular Tumor Board empowerment in clinical oncology practice: a position Paper on behalf of the AIOM- SIAPEC/IAP-SIBioC-SIC-SIF-SIGU-SIRM Italian Scientific Societies. Crit Rev Oncol Hematol. 2022;169:103567. Available from: https://doi.org/10.1016/j.critrevonc.2021.103567 Epub 2021 Dec 8. PMID: 34896250.

[29] Incorvaia L, Russo A, Cinieri S. The molecular tumor board: a tool for the governance of precision oncology in the real world. Tumori 2021. Available from: https://doi.org/10.1177/03008916211062266 Epub ahead of print. PMID: 34918610.

[30] Russo A, Incorvaia L, Del Re M, Malapelle U, Capoluongo E, Gristina V, et al. The molecular profiling of solid tumors by liquid biopsy: a position paper of the AIOM-SIAPEC-IAP-SIBioC-SIC-SIF Italian Scientific Societies. ESMO Open 2021;6(3):100164. Available from: https://doi.org/10.1016/j.esmoop.2021.100164 Epub 2021 Jun 3. PMID: 34091263.

[31] Russo A, Incorvaia L, Malapelle U, Del Re M, Capoluongo E, Vincenzi B, et al. The tumor-agnostic treatment for patients with solid tumors: a position paper on behalf of the AIOM- SIAPEC/IAP-SIBioC-SIF Italian Scientific Societies. Crit Rev Oncol Hematol. 2021;165:103436. Available from: https://doi.org/10.1016/j.critrevonc.2021.103436 Epub 2021 Aug 8. PMID: 34371157.

[32] La Mantia M, Koyyala VP. The war against coronavirus disease 19 through the eyes of cancer physician: an Italian and Indian young medical oncologist's perspective. Indian J Med Paediatr Oncol 2020;41:305−7.

[33] Passiglia F, Galvano A, Gristina V, et al. Is there any place for PD-1/CTLA-4 inhibitors combination in the first-line treatment of advanced NSCLC?-a trial-level *meta*-analysis in PD-L1 selected subgroups. Transl Lung Cancer Res 2021;10(7):3106−19. Available from: https://doi.org/10.21037/tlcr-21-52.

CHAPTER 9

Future perspectives

L. Incorvaia[1],*, M. La Mantia[1],*, S. Cutaia[1],*, V. Gristina[1], A. Galvano[1], N. Barraco[1], A. Perez[1], G. Badalamenti[1], A. Russo[1] and V. Bazan[2]

[1]Department of Surgical, Oncological, and Oral Sciences, University of Palermo, Palermo, Italy
[2]Department of Biomedicine, Neuroscience and Advanced Diagnostics (Bi.N.D.), University of Palermo, Palermo, Italy

Over the years, thanks to the increase in scientific knowledge in the diagnostic and therapeutic field and the refinement of molecular biology techniques, the management of cancer patients has been completely revolutionized. The foundations of precision cancer medicine were laid on this knowledge, which made it possible to move towards targeted choices thanks to the identification of specific and unique molecular biomarkers that act as drivers for the use of target therapies [1]. The key points on which precision medicine is based are the search for specific biomarkers, liquid biopsy, the molecular biology techniques used, and the creation of molecular tumor boards for the correct management of the information obtained [2]. These are all steps of the same innovative path.

Biomarkers in oncology can be divided into five different categories:

- Risk markers: evaluate the risk of developing cancer in high-risk healthy subjects;
- Prognostic markers: able to stratify patients in different risk classes according to a specific outcome;
- Predictive markers: provide data on the sensibility or resistance of the tumor to a specific therapy;
- Surrogate markers: assess the activity or efficacy of the treatment;
- Diagnostic markers: usually employed in screening programs or supporting diagnostic exams.

In this context, the most involved are the predictive biomarkers in different types of cancer such as EGFR, ALK, ROS1, PD1 / PDL1 in lung cancer, HER2, estrogen, and progesterone receptor in breast cancer, c-KIT in GIST [3].

---

* Incorvaia L, La Mantia M, and Cutaia S. should be considered equally co-first authors.

Liquid Biopsy
DOI: https://doi.org/10.1016/B978-0-12-822703-9.00019-3

© 2023 Elsevier Inc.
All rights reserved.

These are certainly the most common but the new attitude in influencing therapeutic choices is that of not relegating the choice of treatment on the basis of histology but in relation to the biomarker driver expressed by the tumor.

The studies that use this new type of approach are basket trials, an innovative type of study that is based on the presence of a specific genomic alteration, regardless of the histology of the tumor to be used as a therapeutic target. This allows us to offer new therapeutic opportunities even to rare cancers for which there are no specific treatments or in any case the therapeutic opportunities are limited [1].

A new class is therefore added to the categories of biomarkers: tumor-agnostic biomarkers which act as links for agnostic therapies, namely drugs that have a therapeutic indication regardless of the site and histology of the tumor [2].

The first use of an agnostic biomarker in clinical practice was that related to the FDA approval, in May 2018, of pembrolizumab in adult and pediatric patients with unresectable or metastatic solid tumors with high instability of microsatellites (MSI-H) or dMMR [4].

Microsatellites are short repeating sequences of 1−6 nucleotides distributed in the human genome whose errors, which occur frequently during cell replication, are recognized and corrected in normal cells by a mismatch repair (MMR) consisting of four proteins main denominated MLH1 MutL Homolog 1), MSH2 (MutS Homolog 2), MSH6 (MutS Homolog 6) and PMS2 (PostMeiotic Segregation increased 2). In cancer cells, MMR genes can be altered or missing, resulting in the accumulation of errors resulting in microsatellite instability (MSI). The deleterious variants affecting MLH1, MSH2, MSH6, and PMS2 can be germinal and/or somatic [5].

Larotrectinib and Entrectinib were subsequently approved as agnostic therapy for patients with NTRK-positive advanced solid tumors [6,7].

The NTRK1, NTRK2, and NTRK3 genes encode a family of extracellular neurotropin receptors, TRKA, TRKB, and TRKC, capable of binding their respective ligands with high affinity, resulting in homodimerization and activation. The most frequent mutations found in NTRAK receptors are fusions; these were present in no more than 1% of all pediatric and adult cancers mainly in childhood fibrosarcoma; to a greater extent in the case of papillary carcinomas of the thyroid gland; even lower frequencies occur for more common cancers such as those of the colon, lungs, pancreas, biliary tract, head, and neck, kidneys, etc. [8].

These innovations have developed in parallel with advances in genomics and technology. The biological technique that has given a boost is second-generation sequencing known as Next Generation Sequencing (NGS). The NGS allows for analysis in parallel, thanks to the use of specific panels, millions of DNA fragments to be able to sequence the entire genome or specific fragments of interest. Nowadays, thanks to the possibility of using targeted therapies, having detailed genomic information in the shortest possible time are increasingly necessary [9].

Today, therefore, the treatment decision is increasingly dependent on the molecular characterization of the tumor.

However, one of the problems encountered in this context is related to the characteristics of the histological tissues analyzed, which are the exclusive representative of a specific space-time interval. However, we know that the characteristics of tumors change, and it is not always possible to have new histological samples: thus, liquid biopsy enters the field [10].

The liquid biopsy allows, using biological fluids as a surrogate for the neoplastic tissue, to obtain useful information for diagnostic, prognostic, and predictive purposes of response to therapies; the most frequently used sample is blood. It has several advantages over tissue biopsy as it is a non-invasive and low-cost procedure, it is repeatable over time, it can represent the molecular heterogeneity of the disease containing, at least potentially, tumor DNA from different areas of the same tumor and different possible locations of the disease [11].

The biomarkers that can be identified and studied by liquid biopsy are very many, primarily among all circulating nucleic acids (CNA), such as circulating tumor DNA (ctDNA) and circulating microRNA (miRNA), but also circulating tumor cells (CTC), exosomes, etc. An example of the use of liquid biopsy in oncology is in the management of patients with EGFR-mutated NSCLC treated with tyrosine kinase inhibitors [12,13].

Cancer management, therefore, becomes more and more complex and personalized, and to ensure the best necessary care becomes necessary a personalized approach between professionals, who must be complementary and have different perspectives. From these assumptions, the idea of the Tumor Board was born, a tool that allows, through the comparison and discussion of complex cases, to define the correct therapeutic path towards which to direct the patient [14].

Nowadays even more useful is the presence of the Molecular Tumor Board (MTB) which, thanks to the presence of a molecular biologist, aims to bridge the gap between knowledge on cancer genomics and daily clinical practice. It is therefore not enough to define a molecular driver, but it is necessary to know the possibilities in clinical practice [15].

The professional figures involved in the MTB are generally the oncologist, the hematologist, the anatomopathological, the molecular biologist, the clinical biologist, the geneticist, the surgeon, and the bioinformatician [2,16−18]. This becomes essential for good clinical practice.

This book and these ideas therefore lay the foundations for identifying an innovative method for patient management by involving as many professionals as possible, each with their skills and being able to overcome, thanks to the help of technology, even geographical distances through teleconsultation.

Acknowledgments

M. La Mantia, S. Cutaia, and V. Gristina contributed to the current work under the Doctoral Program in Experimental Oncology and Surgery, University of Palermo.

References

[1] Goossens N, Nakagawa S, Sun X, Hoshida Y. Cancer biomarker discovery and validation. Transl Cancer Res 2015;4(3):256−69. Available from: https://doi.org/10.3978/j.issn.2218-676X.2015.06.04.

[2] Russo A, Incorvaia L, Del Re M, Malapelle U, Capoluongo E, Gristina V, et al. The molecular profiling of solid tumors by liquid biopsy: a position paper of the AIOM-SIAPEC-IAP-SIBioC-SIC-SIF Italian Scientific Societies. ESMO Open 2021;6 (3):100164. Available from: https://doi.org/10.1016/j.esmoop.2021.100164 Epub 2021 Jun 3. PMID: 34091263.

[3] Russo A, Peeters M, Incorvaia L, editors. Practical medical oncology. Christian Rolfo.

[4] Wookey V, Grothey A. Update on the role of pembrolizumab in patients with unresectable or metastatic colorectal cancer. Ther Adv Gastroenterol 2021;14. Available from: https://doi.org/10.1177/17562848211024460 17562848211024460.

[5] Sinicrope FA, Sargent DJ. Molecular pathways: microsatellite instability in colorectal cancer: prognostic, predictive, and therapeutic implications. Clin Cancer Res 2012;18:1506−12.

[6] Demetri GD, Paz-Ares L, Farago AF, Liu SV, Chawla SP, Tosi D, et al. Efficacy and safety of entrectinib in patients with NTRK fusion-positive tumours: pooled analysis of STARTRK-2, STARTRK-1, and ALKA-372-001. Ann Oncol 2018;29:ix175. Available from: https://doi.org/10.1093/annonc/mdy483.003.

[7] Laetsch TW, DuBois SG, Mascarenhas L, Turpin B, Federman N, Albert CM, et al. Larotrectinib for paediatric solid tumours harbouring NTRK gene fusions: phase 1 results from a multi- centre, open-label, phase 1/2 study. Lancet Oncol 2018.

[8] Cocco E, Scaltriti M, Drilon A. NTRK fusion-positive cancers and TRK inhibitor therapy. Nat Rev Clin Oncol 2018;15(12):731−47. Available from: https://doi.org/10.1038/s41571-018-0113-0.

[9] Kamps R, Brandão RD, Bosch BJ, et al. Next-generation sequencing in oncology: genetic diagnosis, risk prediction and cancer classification. Int J Mol Sci 2017;18 (2):308. Available from: https://doi.org/10.3390/ijms18020308 Published 2017 Jan 31.

[10] Russo A, Giordano A, et al. Liquid biopsy in cancer patients: the hand lens to investigate tumor evolution. Current Clinical Pathology. Springer International Publishing; 2017.

[11] Tellez-Gabriel M, Knutsen E, et al. Current status of circulating tumor cells, circulating tumor DNA, and exosomes in breast cancer liquid biopsies. Int J Mol Sci 2020;21(24):9457.

[12] Wang Z, Yang JJ, Huang J, et al. Lung adenocarcinoma harboring EGFR T790M and in trans C797S responds to combination ther- apy of first- and third-generation EGFR TKIs and shifts allelic configuration at resistance. J Thorac Oncol 2017;12:1723−7.

[13] Tsimberidou AM, Fountzilas E, Nikanjam M, Kurzrock R. Review of precision cancer medicine: evolution of the treatment paradigm Cancer Treat Rev 2020;86:102019ISSN 0305−7372. Available from: https://doi.org/10.1016/j.ctrv.2020.102019. Available from: https://www.sciencedirect.com/science/article/pii/S0305737220300578.

[14] Fennell ML, Prabhu Das I, Clauser S, Petrelli N, Salner A. The organization of multidisciplinary care teams: modeling internal and external influences on cancer care quality. JNCI Monogr 2010;2010(40):72−80. Available from: https://doi.org/10.1093/jncimonographs/lgq010.

[15] van der Velden DL, van Herpen CML, van Laarhoven HWM, Smit EF, Groen HJM, Willems SM, et al. Molecular tumor boards: current practice and future needs. Ann Oncol 2017;28(12):3070−5. Available from: https://doi.org/10.1093/annonc/mdx528.

[16] Schwaederle M, Parker BA, Schwab RB, Fanta PT, Boles SG, Daniels GA, et al. Molecular tumor board: the University of California San Diego Moores cancer center experience. Oncologist 2014;19(6):631−6. Available from: https://doi.org/10.1634/theoncol-ogist.2013-0405.

[17] Russo A, Incorvaia L, Capoluongo E, Tagliaferri P, Galvano A, Del Re M, et al. The challenge of the molecular tumor board empowerment in clinical oncology practice: a position paper on behalf of the AIOM- SIAPEC/IAP-SIBioC-SIC-SIF-SIGU-SIRM Italian Scientific Societies. Crit Rev Oncol Hematol 2022;169:103567. Available from: https://doi.org/10.1016/j.critrevonc.2021.103567 Epub 2021 Dec 8. PMID: 34896250.

[18] Incorvaia L, Russo A, Cinieri S. The molecular tumor board: a tool for the governance of precision oncology in the real world. Tumori 2021. Available from: https://doi.org/10.1177/03008916211062266 Epub ahead of print. PMID: 34918610.

Glossary

Liquid biopsy Components from biological samples (blood, urine, CSF, and saliva may furnish information on the immunological and molecular characteristics of the tumor)

Gene amplification Number of copies of a gene that is increased in certain cells and regions of chromosomes because extra copies of DNA are made in response to cell development signals or environmental stress, with subsequent increase of mRNA and its proteins.

Copy number variations (somatic) A change in the DNA copy number that involves tissue cells, including cancer cells. The alterations comprise loss or gain of chromosome segments or complete chromosome arms and amplifications or focal deletions.

Circulating tumor cells (CTC) Cells that detached from the tumor that go into a biological fluid like blood.

Circulating tumor DNA (ctDNA) Circulating Tumor DNA (ctDNA) is a tumor-derived fragmented DNA present in the bloodstream and many other different biological fluids

Oncogene addiction The tumors growth and survival depend on a single mutated oncogene, thus a selective inhibition of a determined oncogene negatively affects tumor growth and progression

Oncogenes Genes cause the tumor transformation. Proto-oncogenes are regularly present as a nonpathological variant in all the cells, whereas their activation is due to environmental exposition

Extracellular vesicles Small cell-derived vesicles are released into different biological fluids; they can be isolated to obtain information on the tumor's biological features.

Next-Generation Sequencing NGS is a highly sophisticated technique and requires technological acumen and high bioinformatic support, while sangers sequencing is less technically demanding and takes a longer duration than NGS. Sanger sequencing explores strands of DNA to detect mutations by analyzing contiguous sequencing reads.

Real-Time PCR Amplifies the target sequence in the source material by binding with primers and detecting DNA and RNA sequences. Standard PCR includes amplification of a chosen DNA sequence to create millions of copies and empower detection and analysis. The requirement is the target sequence should be known and primers should be available. PCR can be qualitative which answers yes or no for the presence of specific genetic aberration or quantitative which gives titers or allelic fractions of mutations.

Digital Droplet PCR In opposition to classical real-time PCR where amplification is completed in a single reaction volume, in ddPCR reaction combination is divided into thousands of tiny reaction cavities for individual PCR runs. By counting each cavity and determining whether PCR amplification has taken place or not, utter copy numbers of target DNA could be estimated. Using thousands of droplets on a nanoliter scale is a bending and a rather cost-efficient version of ddPCR, defined as droplet digital PCR (ddPCR). The most famous system for ddPCR is Bio-Rad's QX system. This above-mentioned technique is used to identify genetic mutations in ctDNA with high sensitivity and specificity. Circulating cell-free DNA (cfDNA), which includes

DNA derived from tumors (ctDNA), is detected in the blood plasma of cancer patients. The only limit of this method is the small range of targeted mutations detected per assay, while the cost is minimal.

Whole Exome Sequencing A method that examines whole exomes (coding regions). using hybridization, detect the gene of interest, followed by NGS and specialized bioinformatics analysis. Unfortunately, WES is expensive, and also requires more cancer tissue specimens and proper analysis with a lot more bioinformatic support.

Whole Genome Sequencing Examines genomic DNA regions, followed by NGS and specialized bioinformatics analysis. These tests interrogate both coding and noncoding regions of the human genome in a more comprehensive way. The clinical implication of most genomic variations identified by WGS is unfortunately unknown. Non-tumor-specific germline alterations related to disease might be detected as well, generating ethical concerns regarding the correct consideration and significance of these incidental findings.

Multiplex ligation-dependent probe amplification A complex and multiplex assay to identify copy number alterations of genomic DNA sequences. In MLPA a sample hybridized to the target sequence is amplified, regularly within an exon range. This probe is synthesized as two half samples, in the 5′ and the 3′. Additionally, they include universal primer sites and stuffer sequences. With MLPA, more than 40 various genomic sequences could be analyzed concurrently.

Fluorescence in situ hybridization The basic elements of FISH are a DNA Probe and a Target Sequence. The principle is based on the property of double-stranded DNA to denature on heating to form single-stranded DNA. On cooling, single-stranded DNA reanneals with its complementary sequence to reform double-stranded DNA.

Index

Note: Page numbers followed by "*f*" and "*t*" refer to figures and tables, respectively.